THE

ABCQ

OF CONCEIVING

CONCEPTION

Francis Etheredge

En Route Books and Media, LLC

St. Louis, MO

En Route Books and Media, LLC
5705 Rhodes Avenue
St. Louis, MO 63109

Cover credit: Sebastian Mahfood
Copyright © 2022 Francis Etheredge

ISBN-13: 978-1-956715-09-5
Library of Congress Control Number: 2021952733

CONTENTS

Acknowledgments and Blurb... v

Biography and Foreword by John C. Lemon: A Practicable Response to Human Conception vii

What is in a Title? The ABCQ of Conceiving Conception: *The ABC of Conception: An Unfolding Content; The Q of Conception; This Book: The ABCQ of Conceiving Conception; Three Parts*.. 1

Part I

Practical Experience: Comprising Chapters One and Two 23

Chapter One: An Imaginative use of Gardening and Plant Life: *Tomato Seeds, Plants and Conception; Conception and Growing Potatoes; The Integrity of Human Being; Plant Loss and Human Sorrow*....................................... 27

Chapter Two: Gardening Continues to Help us to Understand Conception: *The Possibility of a Single Answer to*

A Unitary Beginning of One or Many

i

When Did I begin? Embryology: What is One Organism?
Reverting to What is Original 43

Part II

Literary Truths and the Literal Truth: Comprising
Chapters Three and Four 57

Chapter Three: Passing Through the Past to the Present:
   From the "literal" use of an Image to the Truth of Em-
   bryology; Stage One: Taking the Comparison with a
   Plant to be Literally True; Stage Two: Movement, Sen-
   sation and the Rearing of Young; Stage Three: Rational
   Ensoulment; A Concluding Reflection: Towards Under-
   standing Human Ensoulment; The Greatest Natural
   Transformation: The Unfolding of Conception 61

Chapter Four: Scripture and Theology: Word and Dogma:
   The word of God and dogma; A Variety of Witnesses to
   Human Conception – Beginning with Eve; Job; David;
   the Martyred Mother of Her Seven Martyred Sons;
   Mary: The Dogma of the Immaculate Conception and
   Human Conception 75

Part III

What is Certain and What is Uncertain about Conception: Comprising Chapters Five, Six and Seven.......... 93

Chapter Five: The Teaching of the Church and the Problem of Uncertainty: *Prologue: A Modern Moment; Introduction: Who is My Neighbour; The Problem of Uncertainty in both Church Teaching and the 14ᵗʰ Amendment*
.... *101*

Chapter Six: On the Interpretation of Texts: Particularly "Amendment 14": *Amendment 14; Mr. Justice Rehnquist, Dissenting; On the Question of the Rightful Protection of Women; and The principle of Determining an Appropriate Level of Legal Action* 119

Chapter Seven: An Answer to the Uncertainty of What or Who Exists at Conception: *What is the Experience of Women in Pregnancy?; The Witness of Each One of Us; A Discussion on the Teachings of the Catholic Church and the "Opinion of the Court"; A Clarification as Regards the Teaching of St. Thomas Aquinas; The Contribution of*

Revelation and Dogma; A Variety of Bioethical Declarations; Gravitating to a Consensus 139

Conclusion:

Lest we Forget Mother, Child and Father.................. 165

Postscript: Roe v Wade: The Ongoing Arguments of Benefit to us All: Justice Beyond a Change of Justices (i); Viability is for Life (ii); Choice, Burdens, and their Alleviation (iii); Bodily Integrity, Liberty, Equality and the Constitution (iv); Brain Death and Abortion (v); Abortion and the Advancement of Women (vi); True Justice is Irreversible (vii). 171

Further Reading: A Variety of Prior Work on Conception 197

ACKNOWLEDGEMENTS AND BLURB

I wish to thank the many people who, over the years, have helped me to develop an account of the beginning of life that is both more widely accessible and draws on both our modern understanding of embryology and a trail of truth in the biblical and historical development of Christian thought.

My thanks to John C. Lemon for his encouraging Foreword at a very early stage of this book, for Dr. Anthony Williams proofreading and for Dr. Sebastian Mahfood's publishing expertise.

I wish to thank, too, the following people who, having read earlier versions of this text and have made a number of helpful suggestions for its improvement: Mr. Michal Paszkiewicz, Dr. Dr. Ralph Weimann, with whom I went over the opening pages repeatedly, and the Rev. Dr. Gareth Leyshon. However, any imperfections and final decisions as to the book's general argument and form do not necessarily reflect the views of those who have commented on the earlier versions and, therefore, the final version remains my responsibility.

The Publisher's Blurb

Francis Etheredge explains what is probably one of the greatest transformations in the whole of nature: the changing outward expression of human development that shows the identity of the person from conception. In these pages, Francis seeks to use what is familiar, ancient, modern, scientific, experiential, personal, and philosophical to help us to appreciate the whole gift of God in the mystery of human conception. At the same time, he appeals to ordinary experience and the human tendency to "image" what we know – allowing him to draw on Scripture to help understand conception and to communicate what we have understood to others. Just as relationships are for the good of all, we can recognize that the completion of human development is a human right for all who are conceived – excluding no one and accepting everyone.

BIOGRAPHY AND FOREWORD
BY JOHN C. LEMON

John C. Lemon is a former pastor of 14 years, TV panelist, founder of John C. Lemon Media LLC, and host of The Podcast with John C. Lemon -- presenting policy, project, publication, and research conversations. Mr. Lemon is a student of cultural systems, corollary strategies and through lines to theology. Interviewees include authors, executive coaches, policy advocates, professors, researchers, social scientists, and university deans. His podcast can be heard, and commentaries read at jcltalk.com.

Mr. Lemon has earned degrees from East Carolina University in Child & Family Relations/Psychology (B.S.), The University of North Carolina at Greensboro in Media Studies (B.A.) and Liberty Theological Seminary and Graduate School -- completing course requirements in missiology and anthropology (M.A. in Divinity).

FOREWORD
A PRACTICABLE RESPONSE
TO HUMAN CONCEPTION

COVID-19 protocols meant working from home and an uptick in logged hours in front of my computer. My productive hours were spent news gathering and cross-referencing guest biographies for upcoming podcast interviews. In the process, I became acquainted with Francis Etheredge through a professional networking site. He was gracious from the beginning – proudly speaking of his surviving eight children and the unfortunate loss of three. As a former pastor, I have walked alongside congregants in search of meaning during difficult times and share with Francis the heartbreak of losing a child.

I am reminded in such moments that our most organic pursuit is to make sense of our world. The essential characteristic of systematic theology, defined as the endeavor to understand and interpret God's actions, runs parallel to that process. Where we land, of course, has all to do with the tools available to us. It stands to reason that in any process, transparency would lend itself to common ground and perhaps adoption. Under that conviction,

Francis acquaints the reader with various fields in search of a practicable response to human conception. His appeal begins in the historical scientific precedent of observable phenomena.

Gregor Mendel's mid-1800's pea plant experiments, garnered the Austrian monk posthumous recognition as the progenitor of modern genetics. Mendel's break-through work made possible the accurate tracing of varied hereditary traits within an organism. Francis follows suit with the life cycle of the potato -- using relatable truths, by way of analogy to explain the mystery of childbirth. Once again, through the familiarity of gardening we are brought alongside to witness the developmental process of human life and, in this moment, reminded that, if unimpeded, he or she will continue to maturity. In so doing, Francis offers a contribution to the continuing question of viability. At what point after conception is that which is conceived considered a person? The triumvirate of embryology, philosophy, and theology is consulted in the belief that truth will not controvert itself. In what is ultimately a plea for national and international bio-ethical policy, we are urged to reflect in deference to ancient insight. We are "fearfully and wonderfully made" (Psalm 139: 14, NIV).

Science is not in any way exempt from this challenge to consciousness regardless of its occasional indifferent persona. Philosophical prejudice lies not only in what we embrace, but also in the principles we elect to ignore. The considered abortion exception of life-threatening harm as rule for the weight of all terminated pregnancy cannot exist apart from bias. The effort here is not to present God without mercy and compassion towards those experiencing the far less publicized deep regret and trauma associated with aborting a child. On the contrary, God is near to heal out of the abundance of His immeasurable goodness.

I was privileged to have a first read of *The ABCQ of Conceiving Conception* and honored to contribute in this small way. Francis demonstrates in this yet pivotal conversation that the question of viability is ongoing. I appreciate and expect that you will as well his labor in raising issues with what some have considered a settled debate.

John C. Lemon
Host of The Podcast with John C. Lemon/Apologetics
Writer, John C. Lemon Media LLC
www.jcltalk.com

WHAT IS IN A TITLE?
THE ABCQ OF CONCEIVING CONCEPTION

What doesn't begin to exist, cannot continue, never mind come to an end. What has an end, has a beginning. What has a beginning, has begun. What has already begun, had a beginning. Whichever or whatever way we look at it, to exist is different from non-existence; and to begin to exist is as different from non-existence as it is possible to be. Just, then, as human existence is a witness to having had a beginning, why is it difficult to accept that another person's existence is a witness to his or her beginning?

A human right is not a right if it is only for you and not for another. What is for one and not another is not a right but an act of discrimination. An act of discrimination is an injustice. An injustice implies justice. Therefore, ending discrimination against the unborn is a right requiring a just resolution.

A right to privacy cannot be a right which implies protecting a wrong. Just as there is no right to deprive an innocent person of life, so there is no right to privacy if it is a cloak under which to wrong another[1]. Therefore, there is no human right to abortion.

---

[1] This was a right claimed to be integral to permitting abortion; cf. Cf. Tina Beattie, "Catholicism, Choice and Consciousness: A Feminist Theological Perspective on Abortion", *International Journal of Public Theology* 4 (2010) 51–75: https://www.academia.edu/18697950/Catholicism_Choice_and_Consciousness_A_Feminist_Theological_Perspective_on_Abortion?email_work_card=view-paper; on p. 62:

'However, liberal arguments on abortion tend to be informed by this post-Enlightenment concept of the autonomous individual, as was evidenced in the appeal to the constitutional right to privacy in the Roe v. Wade case; a landmark case in the United States that made legal history in the abortion debate, but continues to generate considerable controversy.' So although I agree with her sense of how individualism has corrupted the sense of relationship which she goes on to emphasize, nevertheless, Beattie's relationship between mother (and no mention of a father) is not founded ontologically but in some kind of subsequent recognition by the mother of her relationship to the child (p. 66 etc.) and, as such, is as arbitrary as her other claims e.g. 'no very first instant' (p. 60).

Cf. also the following, very comprehensive grasp of the present situation and its historical antecedents and I only choose one of many points it makes: 'The European Court of Human

Imagery is basic to human understanding; and, it could be argued, finding adequate imagery is also an indication of whether or not we have understood our subject and can therefore communicate it easily. Just, then, as moral norms are as integral to human nature as heat is to the flame that expresses it - so the origin of a person's life is when a divine spark ignites it! 2 This book, then, like a catechesis, hopes that it will resound in your heart and that you will find your own way to communicate the beginning of each human life to this generation.

Rights has held that a mother's right to privacy cannot be interpreted as a right to abortion': "Catholic Medical Association (UK) Submission to the World Medical Association Consultation Regarding the Revised Draft of the International Code of Medical Ethics": http://www.cmq.org.uk/CMQ/2021/Aug/ Submission_to_WMA_re_conscience.html.

2 Cf. St. Paul VI, *Humanae Vitae* (On Human life), 13. Note: A document of the *Catholic Church* takes its title from the opening words of the Latin text and, therefore, the English translation only approximates, in some cases, the original meaning of the Latin.

This book, then, takes a closer look at the growth of plants with a view to helping us to understand human conception. In other words, taking what is more within our immediate experience, we will consider what helps us to understand human conception. Thus, this is not an unusual activity; St. Thomas Aquinas says that it is helpful to go from what is familiar to what is unfamiliar; and, as we might imagine, this principle has been used from ancient times. Although the biblical evidence is a different kind of source to biological knowledge, it makes use of what is familiar to help us to understand what is unfamiliar. Thus the author of the book of Job compares his origin with the curdling of cheese: 'Didst thou not pour me out like milk and curdle me like cheese' (Job 10: 10); and David says that God made him an 'unfinished vessel' (Psalm, 139: 16), signifying both the possibility of growth and of being 'earthen vessels' that show forth the power of God (2 Cor 4: 7). But just as Eve has already acknowledged the creative action of God at conception when she said: 'I have gotten a man with the help of God' (Gn 4: 1), so the dogma of the *Immaculate Conception* of Mary helps us to understand the wholeness of the human being at conception –

because Mary cannot be wholly holy if body and soul are not one from the first instant of fertilization[3].

At the same time, drawing on the available scientific knowledge of human conception, it can be said that there is a real beginning of the human embryo from the "moment" that the sperm enters the wall of the ovum, and the once inert ovum is now activated, closing around the sperm head and proceeding on an uninterrupted trajectory of human development. Thus, there are simple truths, namely that conception is the beginning of a new human life; and, in the case of all who have ever existed – each and every one of us has had a beginning. Indeed, the very ongoing life that we live is a living witness to the fact of our beginning. Our development has passed through a number of watersheds: the formation of the human embryo from the union of the father's seed and the mother's egg; implantation in the mother's womb; ongoing development; birth and the constant growth of the child.

The transmission of human life from one generation to another grounds the recognition that each one of us is equally given the gift of being a person; and, therefore, co-extensive with being a person is being a person in a

---

[3] This will be more fully explained later in the book.

community of persons: the human race. The rights, then, that pertain to human beings, pertain to human beings from conception; for these rights are integral to the relationship between one human being and another. Human rights, therefore, are inter-relational and express our common humanity: a common humanity which makes each of us a specific human person in the community of peoples.

Finally, then, towards the end of this book, we encounter the social implications of a clearer understanding of human conception: that it can lead to a recognition of the necessity for a widespread, indeed international bioethical law for the benefit of all mankind. Indeed, there needs to be a law that embodies the multiple rights of the nascent human person: the right to life of every human embryo; the right to completing human development; and the right of ensuring the human integrity of each human being, so that nobody's human identity is compromised by being "mixed" with animal ingredients. Thus, all nascent human beings need to be protected from being "mixed" with "ingredients" from other species, being frozen, experimented upon or discarded. In sum everyone has a right to the integrity of his or her identity and human dignity made as we are 'in the image and likeness of God' (Gen 1: 16f)'[4].

The ABC of Conception: An unfolding content

As some have argued that a person does not necessarily exist from the beginning of human conception, it is necessary to discuss the original meaning of conception. Because we have not always existed, our existence has a beginning: '[a] beginning' is the archaic meaning of "conception"[5]; and conception is also understood to mean "to take in and hold; become pregnant"[6]. Ordinarily, then, we understand by conception that the man's sperm has been received and retained by the woman's egg and that she is

---

[4] John O'Brien, OFM, *Waiting for God: From Trauma to Healing*, Poland: Printed by Amazon, 2015.

[5] Con-ception means '[a] beginning' (*The American Heritage College Dictionary*, senior Lexicographer, David A. Jost), p. 228: con-cep-tion: the archaic sense of which is 'A beginning ...'.

[6] Cf. "Conception" (noun): https://www.etymonline.com/word/conception; and cf. also "to conceive" (verb): https://www.etymonline.com/word/conceive.

"with child". Conception is therefore through the husband and wife's gift of themselves to each other and, just
as love goes out to others, so beyond the gift of this self-
giving is the possibility of a child. But just as God regards
the good of human life as distinct from the fallen nature
of man through which we come to exist (cf. Gn 3-4), so
God regards the good of human life even if the child is
conceived through rape, in a glass dish or by any other
means than through the spousal union of husband and
wife. If, then, the father's gift of what transmits human
life, his sperm or seed, becomes one with the mother's gift
of giving the ready-to-receive "egg", then conception is
expressed in the moment of the egg enclosing the sperm.
More precisely, then, the first instant of fertilization is the
first instant that the sperm or seed of the man animates
the ovum – because up to this point the mitochondria, the
energy-producing cells in the ovum, are inactive. In other
words, the closing of what is now the embryonic wall is
the first act of a totally new human being. Thus, just as a
seed is germinated and planted, so is the life of the embryonic child, a newly formed and independently growing,
indwelling person, both con-ceived within the mother
and "im-planted" in her womb.

On the one hand, the inheritance of our common humanity is through the reciprocal self-giving of the parents; but, on the other hand, the nature of being a person exceeds what is transmitted by the parents. Human development, then, shows forth the implicit dynamism of the person conceived. Beginning, then, with whether the child is male or female, development unfolds all that is entailed in being the person each of us is conceived to be. The natural unfolding of the child's humanity is expressed in the physical, psychological and social development characteristic of each one of us. Our human being is inseparably physical, social and psychological; and, therefore, as the child is conceived so begins, in that same instant, the integral being of the human person: each of whom is a being-in-relationship. There is an implicit social nature to the human embryo in that not only does he or she come to exist in virtue of what she or he has received from the mother and the father – but also because the very presence of the child in the womb expresses a particular relationship to the mother. There is both a kind of biological dialogue in which there are subtle exchanges between the mother and the child; and, as the mother's awareness of the child grows, literally, with the growing of the child, so the mother-baby dialogue is also psychological. Let us note

Elizabeth's awareness of John the Baptist's response to
Mary being pregnant with Jesus (cf. Lk 1: 44).

The child exists, at once in the present, in relationship
to parents, relatives and any number of other people. Indeed, as the mother looks forward to the birth of her child,
so she looks forward to meeting the child she nurtures;
and, similarly, as the father witnesses the unfolding of his
child's development, so he begins, too, to look forward to
the birth of their child. The embryonic unfolding entails
psychological development and, therefore, as the child responds to the voice of the parents and, from birth, to being
fed and to smiling and to amusing sounds, so the whole
world of relationships multiplies with grandparents and
the many people with whom they all come into contact
and with whom they are all connected.

As the child begins to put into words his or her experience of life, however simply, so we can see that our inner-
life, our consciousness of our own experience of life, develops through our communication with each other. Indeed, we can even call it a law of psychological development that what is expressed contributes to our human development and what is inhibited or repressed obstructs it:
communication of the self enables the differentiation of

the self through the relationships that help us to know ourselves through knowing each other.

Just, then, as there are the diverse laws of physical, psychological and spiritual development, so the soul is the principle of unity in the human being; and just as laws regulate the fulfilment of coming to exist, and exist as inner determinants of growth, so there is the Divine Lawgiver who brings the whole person to exist, one in body and soul. In other words, the invisible act of God which brings to exist the embodied human person, one in body and soul[7], is shown forth through all that unfolds from conception. Physical and psychological development, interconnected as they are, shows forth the primordial unity of the person: that the nature of the person transcends matter in that the human body expresses a human soul[8]; and, at the same time, as the relationship to the parents develops, so the spiritual nature of the whole person who

---

[7] Cf. *Gaudium et Spes*, (Joy and Hope) 14.

[8] Cf. Pope St. John Paul II, *Familiaris Consortio* (The Family Communion), 11; *Evangelium Vitae* (The Gospel of Life), 60.; and *Veritatis Splendor* (The Splendor of the Truth), 58-50.

exists[9], by virtue of being created by God, is called to a relationship with Him.

The Q of Conception

There are many questions which can surround our understanding of what is going on; indeed, these questions can be like a wall of brambles, a fire wall or a dense fume-fog, which prevent us from recognizing the simple truth of human beginning. Are we justified in describing the father's sperm as seed? Are we justified in describing the mother's ovum as an inert egg? Does the child's life begin before conception? Is the sperm or the egg the beginning of the life of the child? Does the child's life begin after the fact of conception? Does the child's life begin at birth? Does the child's life begin when the child becomes conscious?

The father's sperm is like a seed in that it is a small dynamo, seeking the enclosure which will enable it to energise the growth of a child. However, while the father's seed contributes to the growth of a child, the mother's egg

---

[9] *The Catechism of the Catholic Church*, 28, 44-45, 327, 355, 362-368 etc.

encloses the seed, which has shed what is unnecessary, and enables that seed to drive the changes which transform the immeasurably larger egg into a single celled human child. A child is not the father's seed nor is the child the mother's egg; but a child is the coming together of the father's seed and the mother's egg. Even in the case of twins, which have originated from the same conception, when the early embryo divides and there are two children, then each child has a beginning; one from the beginning of conception and one from the point of being divided from the original embryonic child's development. In other words, where the child's embryonic body lives, there the child is developing. As regards the possibility that, in the case of the splitting of a single embryo, forming two human embryos, there is the death of the first child and the beginning of two more, it presupposes that there can be a death without a dead body. In other words, so integral is the body of a person to being a human person, that there cannot be a human person without a human body; and, therefore, if there were to be the death of a first child, giving rise to a second and a third, then there would have to be the evidence of the death of the first child's bodily expression. In view, then, of twins arising out of the division of a single human embryo, the very continuity of there being two lives gives evidence that one life has continued and a second has arisen from it.

A person might argue that because the human embryo does not look like a mini-adult then the human embryo is not a human person: that what does not look like a human being is not a human being. But does the child look the same at the end of life, if he lives a long life, with greying hairs, perhaps less teeth, scars, maybe glasses and walking stick, wrinkles, moustache, and beard? In other words, even from birth the process of growing up goes on and changes the outward appearance of the boy becoming a man, a man going into middle-age and a middle-aged man becoming older and even elderly. Even as we go from childhood to adolescence there are changes that might make the child unrecognizable as, for example, the young child who is scarcely more than a toddler growing to be-yond six feet tall. Ordinarily, then, growth entails change and change expresses what is constantly present, whether male or female; and, therefore, a man manifests what is there from the beginning and what is there from the be-ginning entails what develops into a man.

But, says a woman, my body is mine? Our bodies ex-press each one of us; and, as such, my body is essential to the gift I have received in being given to be me. I am born, so am I conceived to be born; and, if there is a problem with my development, nevertheless, the orientation of the whole process is to birth. Indeed, in order to accept what is no longer an expression of her own body, the mother's immune system has to let it accept the indwelling of the child; and, therefore, it is clear that the woman bears the child conceived – but bearing a child is a relational process: a relation between mother and child and father. The problem with the perception that 'my body is mine' is that there is no sense of being a gift-in-relationship: of having received a gift of existence and, in a sense, out of gratitude, being willing to contribute to the gift of the existence of another. Just, then, as the seed left my father and the egg was released within my mother, so being conceived is *en route* to being born. The child, then, is not an organ of the mother. The child's placenta, which sits within the womb and, ordinarily, comes away with the birth of the child, helps the mother accommodate the presence of the child; it acts as an "exchange", governing what the mother gives the child and what the child excretes. The placenta, then, is a temporary organ and, as such, is a witness to the child dwelling, temporarily, in the mother; and, indeed, the very growth of the child testifies to the independence of the child from the mother for, ultimately, the growth of the child is towards being born, fully formed, with a view to the ongoing growth and development of his or her life.

What we need are helps to the imagination; for, in general, while many of us will have seen pictures of early embryonic human life, the birth of a child, or babies, there are still many difficulties recognizing what we see as "being one of us".

This book: The ABCQ of Conceiving Conception

We need, then, to draw on a variety of comparisons to help us to understand the mystery of human conception; and, therefore, human conception is put in the context of the natural imagery which helps us to know what is happening. We need help to recognize what we see; and, recognizing what we see entails understanding it. One of the principal helps to understanding is imagining comparisons which make what is unfamiliar more familiar. However, imagination can both take us away from what exists or it can lead us to it; and, indeed, imagination itself tells us about ourselves too.

But if we begin thinking all things are possible and that there is no limit to what we can imagine then we have already begun to envisage possibilities. We can imagine, then, that what we are is something semi-insubstantial, like a computer program, which can be downloaded onto a variety of different computers[10], without altering what it is. While, however, this use of the imagination has worked very well for computers and the programs which run them, what it consists in is a compatibility between the program and the computer: a set of instructions, however adaptable, and a machine that can recognize those instructions and action them. There is, as it were, a hand and glove relationship between the computer and the program. In the case of human beings the person is not a program, "life", and a, "body", that life activates; rather, the human person is that totality of "human life" and "body". While the body lives, there is the life of the person; and, according to the kind of activities that are possible we understand ourselves more clearly.

Thus "imagination" is that by which we envisage many paths, either as goals of understanding or goals of action; and, while we imagine these different possibilities, there is

---

[10] Cf. Stephen Law, "Would You Want To Live Forever":
https://www.linkedin.com/pulse/would-you-want-live-forever-stephen-law-phd/?trackingId=ONAfacRLSuyuKtT5%2FfDzqw%3D%3D.

a dialogue with ourselves and what exists as to which one is really possible for us – either as individuals or as members of the human race. Walking or travelling is about as universal as it is possible to be in that if we are disabled, and cannot walk, we can still travel by various means to a destination; and, equally, there can be a wide variety of reasons why we want to go – all of which involve choices of one kind or another. Choosing goals is both a common characteristic of human beings and distinguishes us, too, by that use of the imagination whereby we bring to exist what did not exist. Recognizing our goals as real and good, our methods too, entails feedback from reality, which enables us to reason to the truth and, if necessary, modifying our goals or methods, and the whole toing and froing which this involves as we go from task to task. Imagination, choice of goals and the use of reason are all factors which express being a person.

The imagination, however, with which this book is concerned, is what we use when we want to understand what is unfamiliar on the basis of what is more familiar to us. No doubt, as this book progresses, it becomes more difficult to understand because it is necessary to draw upon embryology, philosophy and theology to help us; and, therefore, we need to begin simply, remembering that simplicity, when the relationship between law and human rights make it more complicated.

Three parts

What seems, then, to be an unusual and unfamiliar approach to the subject of human conception, of seeing how we have and do conceive of conception, as we will see, is perhaps more characteristic of how people have tried to understand our origin than we may have first realized. But, in the light of modern science, we have become clearer about when and where our natural capacity for making comparisons or analogies can be improved or abandoned for the simple truth of what we now know to be true of our beginning. At the same time, however, we do need to be able to use our imagination to help us to grasp, by comparisons, what is in reality both a visible and an invisible reality: the instant of coming to be a whole human being, one in body and soul.

The book is divided into three parts; and, it is hoped, while these parts address different aspects of the whole, that they are complementary. The book, then, seeks to stimulate the founding of an internationally binding charter which ensures the natural integrity and the completing human development of everyone, from conception until natural death.

Finally, the book cannot be exhaustive, nor is it intended to be; rather, it is hoped to be a start, even if its starting point is not zero; but, nevertheless, it is written with a view to helping a person become more familiar with the ordinary extraordinary event of human conception. Indeed, "conceiving conception", as we shall see, entails making many "comparisons", both historical and in the present, that either help or hinder becoming clearer about the reality to be investigated: the beginning of each one of us. Just, then, as there is an embryological science which sets out to identify what is known about the reality of human beginning there is a need, too, to find ways to connect this with our ordinary experience and so to be able to communicate the truth of a real beginning more and more widely. Thus, the more widely the truth of an actual, real, irreversible beginning to each one of us is capable of being communicated the greater the possibility of discovering a legal expression of who, in the end, exists in relationship to all of us from conception onwards.

PART I

PRACTICAL EXPERIENCE
COMPRISING CHAPTERS ONE AND TWO

This book began, as it were, with the third and more difficult discussion on the need for bioethical law, to both identify and protect the human person from conception onwards; however, the third part opens with the passing observation that a runner bean seed produces a runner bean plant, a tomato seed, a tomato plant and a potato, a potato plant. A child, then, is not an abstract "thing" but is from the first instant of fertilization a child of his or her parents; and, at the same time, if there is no child then there are no parents: a child is always a sign of the relationships between the generations: between who and how the child is generated. In other words, without realizing it, *then,* all the time I have been growing vegetables in the garden I have been thinking about the consistency of what grows. In other words, there is a definite value in our everyday experience and, therefore, I encourage us all to consider the value of the everyday arguments for the first instant of human conception and the bioethical questions which are involved.

The lighting of a candle, for example, can literally ignite the thought that there was an irreversible moment when the wick was not alight and now it is; or, if the light is not passed, then the ignition of the candle has not occurred. But, on the other hand, if the candle is lit and then goes out there is the equivalent of a miscarriage. Alternatively, knowing that there are innumerable mitochondria, or power generating units, present in the woman's inert egg, it is clear that lighting the candle is passing a light to what can sustain it owing to the presence of the body of the candle; so, as with the body of the candle, so it is a mixture of ingredients, so with the embryonic cell body, it is more than the power generating units that will drive its activity. But, nevertheless, without the light being passed, the candle remains inert, just as the egg remains inert and passes away if not fertilized.

Alternatively, however, it could be the imagery that arises out of technological devices, like the sending of the text becoming, as it goes, irreversible and, even if the device is destroyed, the text remains on a server somewhere in the "world"; existence, then, is an irreversible condition of what exists and, therefore, once the person has come into existence, his or her existence is now an existence-in-relationship: a child of parents.

With the experience of gardening, then, there has arisen a whole range of thoughts to do with the beginning of each one of us and related bioethical questions; and, even if not everything is discussed or finds a suitable parallel, the value to be explored, each in their own way, is the value of everyday experience being a vehicle for addressing, in a wider and more accessible way, the challenging questions that arise in human embryology.

CHAPTER ONE
AN IMAGINATIVE USE
OF GARDENING AND PLANT LIFE

We need what is familiar to help us to understand what is unfamiliar; however, we also need what is familiar to be so. Therefore, while in one sense computers and computer speak are more familiar to most people than gardening, gardening needs to become more familiar again to help people to understand the action of nature in the growth of plants and, ultimately, human beings. So this chapter is both advocating a return to sowing and growing plants from seeds and the use of that imagery and experience to help us to understand the mystery of human life. Different plants, as we shall see, are helpful in different ways; but, as a whole, what is sown is what is grown: a tomato seed or a seed potato grow into tomato and potato plants. If, then, a human being is conceived then nature will unfold the person who is present to maturity.

In one crucial respect, however, human beings and plants are different. Although plants can grow wild, many vegetable type plants are cultivated and planted by human beings; and, in certain sense, this "husbandry", or care of

the household[11], means that many more plants and vegetables are planted and cared for and provide beauty and
nutrients for a family. In the case of human beings, however, the complementary contribution of the "gardener"
who plants and waters is God who, on the coming together of husband and wife, brings about the existence of
the soul embodied in the body. Just, then, as the seed of
the plant receives favorable conditions for its growth from
the farmer, so God makes possible the unique wholeness
of the human person by animating the parental ingredients with an embodied human soul; indeed, a human soul
cannot but come to exist as the soul of a specific person
and, therefore, the instant that a particular human being
begins from the contact of "seed" and "egg" then God
gives the life of the bodily person.

Tomato Seeds, Plants and Conception

A seed is by definition the almost ready-to-grow start
to a plant; and, as such, seeds come in various sizes. More

---

[11] "Husbandry": https://www.merriam-webster.com/dic-
tionary/husbandry.

technically, a seed is an embryonic plant in a case[12]; and, when it is dried, it is inert or dormant: it is unable to grow into the plant it is. Tomato plants grow into small bushes unless their branches are clipped off to stop them spreading wildly and, after the flower, the tomato forms behind it, swells and turns red. The seeds come from within the flesh of the tomato and, generally, are dried; they are tiny and can be planted one at a time using the moist end of a matchstick to dab the seed up and to push it into the soil. A dried tomato seed is, generally, ready-to-grow, but needs to be put in moist soil, watered and encouraged with sunlight.

We can see, then, that it is natural for the seed of the tomato to be formed within the tomato itself; the seed is formed for the next generation of tomato plants. The reproduction of a plant, how the next generation will grow from the last one, is clearly a part of how nature functions. However, the tomato seed is clearly different to the rest of the tomato; indeed, while it can be eaten, the seed itself is what will bring about the next generation of tomato plants. What is noticeable, too, is that in a way, the fleshly part of the tomato is also food for the tomato seed, should

---

[12] "Seed": https://en.wikipedia.org/wiki/Seed.

it fall into the soil and need nourishment suitable to its growth. The tomato seed, then, although it is a natural part of the whole tomato, can also be separated out and sown into the soil; indeed, the seed is both different from the plant and the soil and yet comes from the plant and can take up the nutrients it needs from the soil. Thus a part of the preparation for growing the seed is nourishing the soil with compost, rotted or rotting vegetables and manure, faeces from horses, cows or donkeys. The growth of the tomato plant goes on quietly in the dark of the soil and begins to show itself with the first two leaves coming up from the seed and the roots going down. The roots are not normally visible except when the plant is being moved or it is in a glass or transparent plastic container. The plant, then, is capable of being transplanted and, there-fore, if the little plant begins indoors, it can both be put in a bigger plant pot as it grows and, eventually, it can be planted in a bucket or the ground, ready for when it will mature and start to produce tomatoes.

Similarly, the "seed" of the husband and the "egg" of the wife are formed in the course of their ordinary every-day growth from embryo to adult. The "seed" of the hus-band is not passive, like the tomato seed; it is living and moves and, once passed to the woman, seeks out the

"egg". The "egg" is not closed, like a boiled egg, but is rather more like a large ball of wool, which is a mass of openings in which a sperm, the male seed, can come to rest in
one of them. While, however, the seed of the husband is
sown in the body of his wife, the wife's "egg" is released
internally and, if fertilized, then a child is conceived and
will be naturally transported to the mother's womb to im-
plant and to continue to grow. The husband's seed is very
small and travels, as it were, to meet his wife's "egg" which
is comparatively large and, not only does it contribute to
the child's genetic inheritance but, in reality, from fertilization onwards, is substantially the body of the child from
which all development subsequently proceeds. Thus the
husband's seed is like a cotton thread in comparison to his
wife's "ball of wool"; or, alternatively, the sperm is like a
key which turns in a lock to open it. But these inanimate
comparisons do not do justice to the organic structure of
sperm and egg. The woman's "egg" has pores which, once
they enclose a specific sperm from the husband, naturally
close and encase it; hence, the woman's ovum or egg is
both a receptacle of the husband's seed but also the foundational "flesh" of the child's body.

As with the soil, the general health of the mother and
the father is a part of their preparation for conceiving a

child; and, in the case of the mother, for bearing a child although, unfortunately, she may not be as well prepared as she might be.

At ovulation, the mother's egg is released and is carried down an internal pathway to meet the coming of the father's sperm, which already indicates that the child, once conceived, is independent of the mother; not independent in the sense that the child does not need her maternal love, but independent in the sense that "seed" and the "egg" form a new being who will continue to grow. Just, then, as the tomato carries the seed within it, so the mother carries the child within her. Indeed, just as drying the tomato seeds prior to planting emphasizes the "separateness" of the tomato seed from the parent tomato plant, so the coming together of the father's seed and the mother's "egg", help us to see that the child, the human embryo, is neither wholly from the mother nor wholly from the father – but equally from both.

The fact that the child lives in the womb of the mother, initially, is not unlike the seed of the tomato plant being first sown in a plant pot until such time as it is sufficiently well started to be planted out in the ground. The main point of comparison being that the soil and the plant are different. Just as the seed is planted in the soil and benefits from it, so the child is conceived in the mother and benefits, immeasurably, from her maternal care. As the plant and the soil are separate, so are mother and child; although, in the case of father, mother and child, their relationship is forever, while in the case of plant and soil, their relationship lasts as long as the plant lives.

Conception and Growing Potatoes

Potatoes, however, while they can be called "seed potatoes" are otherwise known as "tubers" from a Latin word which means edible root[13]. Thus, by contrast with the tomato, which grows behind the flower and hangs off a stalk on the plant, the potato grows underground. Potatoes come in various sizes and have an "eye" from which the new potato plant will grow; indeed, it may well have been the fact that a number of potatoes begin to grow shoots in a dark cupboard, where they are kept for cooking, that

---

13 "Tuber":

https://www.vocabulary.com/dictionary/tuber#:~:text=A%20tuber%20is%20a%20plant,the%20plant%20considered%20a%20tuber.

may have stimulated my interest in them. Potato plants are remarkable for the fact that they will grow from the slightest piece of potato provided, that is, it has an "eye". Generally, however, there is a period of dormancy, or rest, from when the potato has been taken out of the ground, or left in it, and it begins to send out shoots; this period of non-growth, or dormancy, ceases as the ground or cupboard temperature rises and so the potato starts to grow shoots.

When a potato starts to grow in the cupboard the shoot is often very pale and brittle; putting it on the windowsill brings it into the light and helps the plant's shoots to become green and leafy. The roots do not generally appear until the potato has been put in a little soil. Altogether, although the potato can be planted without any obvious growth, it is interesting to watch for the start of these developments and then to plant them; indeed, when multiple eyes have sent shoots out, it is possible to break or cut the potato up, giving each piece a "shooting eye" – but no smaller than a golf-ball. However, I have seen slices of potato, no bigger than a large coin, with a root and stem. Owing to rotting vegetables being what they are, a mixture of peelings of all kinds, it is often possible to find that some slivers of potato are simply growing up through several inches of soil and starting to leaf as they come out into the light. Be careful, when transplanting these plants, because of their fragility.

When the flower on the potato plant has come and gone, and it is time to dig it up and see if there are any potatoes in the ground, look out for the little tubes which run from the main plant to the potatoes themselves. Thus, the potatoes grow in the ground at the end of little tubes which come off the main plant. This is as near to an umbilical cord as I have ever seen in a vegetable; and, in its way, makes the growth of the little potatoes a vivid illustration of what happens with the growth of our own children.

Potatoes are an altogether more substantial start to growing potato plants than seeds and, therefore, in some ways they resemble the more substantial "egg" which comes from the woman; and, just as the potato has to come out of its state of dormancy, as do seeds generally, so does the mother's inert "egg" need to come out of its state of readiness to be fertilized. Fertilization, then, is that act and process whereby what is ready to develop receives what it needs to develop. Whereas in the case of the potato, dormancy indicates a potato resting before it develops into a plant, the inertia of a wife's "egg" is not just dormancy, as in a resting-state, but a resting-state that needs the completing contribution of the husband's seed. For the potato, then, it is a matter of time and temperature before the tuber, the root potato, starts to grow into a potato plant and, in the process, consumes the rest of the potato as its growth nutrients; indeed, if a potato is on the windowsill in this early stage of the plant's development it is possible to see it start to shrink as the plants takes the nutrients for growth. The potato, however, has in some sense what it needs to grow already present within it, as a potato will start to shoot in the dark or on windowsill and can then be put in the soil to feed the changes of growth which bring about a plant. In the case of the inert egg, when it combines with the husband's seed it is released from its inertia and in the same moment it has become an embryonic child. The first visible sign of this change is the encasing of the husband's seed in the mother's "egg" ceasing to be inert and enclosing the sperm head within an embryonic wall or "skin"; and, as the interior activity develops, so the separation between "sperm or seed" breaks down and the activity that began with the contact between the sperm and the egg continues. Thus the presence of the child grows through cell division, consumption of nutrients and the process of implanting in the mother's womb.

The dormancy of the potato and the tomato seed, although different to that of the mother's inert egg, still suggests the same possibility, namely, that learning how nature operates enables the parents to recognize the time when a child is more or less likely to be conceived. Thus, while accepting that a child may be conceived unexpectedly, there is nevertheless an opportunity to space the conception and birth of children rather like a gardener gives room to each plant. At the same time, however, a person is different to a plant and cannot be weeded out once sown, as the conception of a human being enters that child into the rights of the human race: a human being, on conception, has a right to completing human development.

Just as with a potato that has many sprouting "eyes" it can be split into many potato plants, each with an eye, so it is possible that twins and sometimes more children can come from the separation of a single human embryo as he or she develops.

The integrity of human being

Tomato seeds grow into tomato plants and tomato plants grow tomatoes and tomatoes have seeds which can be dried and sown and grown into tomato plants. While it is true that there is a natural integrity to plant, animal and human identity, it is also true that it has long been the custom to cultivate plants and animals in ways that alter them. In the case of human beings, however, because this involves the life of a person, who has a right to being loved as who he or she is, it is not right to treat the human person *as if he or she is a product to be manipulated.* A justified intervention for the sake of the health of the child is one thing; but an "experimental" intervention for the sake of the experimenter is totally different.

On the one hand, then, it is quite clear from human growth that we consume a wide variety of ingredients, whether from animals or plants and that these nutrients, ordinarily, contribute to healthy growth. On the other hand, introducing genetic or reproductive tissue from an animal, or even a plant, either into the moment of conception itself or into the early stages of human development, raises extremely serious questions of what is, unless known to be otherwise, a human being with a mixed human and animal inheritance. The claim that we are all animals, so there is no difference between "us" and "them", and we can mix genetic or reproductive elements regardless of species, ignores what common experience has shown: that many attempts at cross species reproduction are infertile. The number of factors influencing the compatibility of species already makes this mixing of genetic ingredients perilous from the beginning.

Furthermore, however, as the human embryo develops within the mother, it is now more clearly recognized that there is a genetic dialogue, called "epigenetics", which takes place between the child and the mother whereby each subtly influences the other. In a word, then, it is a gross simplification to describe us all as "animals", as if there is no differentiation between species which makes certain combinations "problematic"; and, indeed, so subtle are many of these differentiating features that even a slight change may have a profound effect on human development. Thus it is no justification to say that as a child can be aborted, we can do what we like with it; as, already, an unwarranted decision has been made about our relationship to that child. Each of us is a gift and no one has the right to take that gift from another; indeed, even adults who have committed crimes, it is argued, can be securely detained in the hope of reform, rather than being executed.

In general, then, irrespective of whether a person may think that what he or she is doing does not involve a human life, there is sufficient uncertainty to denote an absolute prohibition on human-animal species experiments. In the end the Creator may well accept that a creature is human, even if it is in fact ambiguously so, precisely because God sees the person that has come to exist, however imperfectly, is still a human person. Just, then, as the Creator ensouls a child conceived out of wedlock, in a test-tube or because of rape, out of love of that child – precisely because the child is not responsible for the imperfect conditions of his or her coming to exist, so the Creator recognizes that what constitutes human being, however imperfectly, calls forth the power of God to ensoul him or her on coming to exist. At the same time, however, the injustice the child suffers on being brought into existence in a certain way is not justified by the good that God brings out of it.

Plant Loss and Human Sorrow

It is true, however, that just as there are various plant diseases and problems, so there are problems with the first stages of human life; indeed, conception does not always occur and, even when it does, the child may die.

This year, I have planted numerous runner bean seeds, a first group were planted in ice-cream boxes in the house

– but all but one did not develop; indeed, when I examined all the ones that did not develop, all of them had turned into a shapeless gooey mess. My mother-in-law, an experienced gardener, thought the seeds had been too cold and thus did not even develop into plants. Similarly, in the case of what is thought to be a pregnancy, there is no child but an amorphous mass of cells. In other words, something can so radically alter what happens after the contact between sperm and egg that there is no transmission of life but, rather, a growth which is no more like a child than the sticky mess of a spoilt, germinating bean, is like a bean plant. The technical name for an amorphous mass, following the contact of sperm and egg, is that it is a hydatidiform mole: a name which comes from two words, 'the Greek word *hydatis* meaning a drop of water, and the Latin word *moles* meaning a mass'[14].

---

[14] There can also be a partial hydatidiform mole, too, which usually ends with the embryo being miscarried and lost: https://link.springer.com/chapter/10.1007/978-3-642-84385-3_18; indeed, it is very difficult to tell if there is a true child present but, nevertheless, if there is a child then there is the grief because of the loss of that child.

In other cases, as with a reasonably robust seedling that is put out too soon and the frost shrivels its leaves and it dies, the child is actually conceived and begun to develop but for whatever reason, sadly, the child is lost through a miscarriage: passing too early from the mother's womb.

In general, then, it is possible to open up the whole subject of life, and human life, through growing vegetables and to help people understand the simple truth that from a tomato seed comes a tomato plant, and from a potato comes a potato plant; and, therefore, from the coming together of man and woman, through the embrace of marriage or the marital embrace, very often comes the fruit of children.

CHAPTER TWO

GARDENING CONTINUES TO HELP US TO UNDERSTAND CONCEPTION

An Origin of one or many Mae.

Each discipline, whether it is embryology, philosophy or theology, has asked the question of when a human life begins and none of them, it seemed, had a uniform answer. This shows, in its own way, that the questions we ask are influenced by a whole range of factors and, at the same time, that there may be a particular difficulty in answering the question itself. Inevitably, then, there is a variety of answers which, down the years, have tended to recur, albeit inconclusively. On the one hand there is the view that the beginning of the human person is after the very first instant of beginning; and, therefore, that the person is present from any time after or from the fusion of the two pronuclei, each of which came from the male seed (or sperm) and the female egg (or, to use its Latin name, ovum). In other words, according to this view, the person is not present from the very first instant of fertilization. On the other hand, the embryological evidence, philosophical reasoning and theological thought influenced by *the word of God*, indicate a simple beginning which is the very first

43

instant of active contact between the sperm and the ovum or egg. In a word, conception means, a beginning, and just as we recognize beginnings in all walks of life, so we recognize that the word has a specific sense. Indeed, if I turn a key in a clock, it begins to take up the mechanism and wind up to an activity; and, after the winding, the mechanism is set in motion. Thus the telling of time begins from the tick-tocking of the clock. Thus there are activities that lead to a beginning but are different to it; and, therefore, after is not the same as before and is different from it. A child, then, comes to exist both because of the union of sperm and egg and the hidden action of God which goes beyond and completes what is visible; and, indeed, this will be discussed later in the book.

The possibility of a single answer to when did I begin?

What makes the possibility of a converging answer possible from these different disciplines? There are two, if not three answers to this question. Firstly, each of them is interested, as it were, in the question of beginning; and, therefore, there is a common "object" of any inquiry which has arisen concerning the beginning of the human person. In other words, the beginning of the human person stands as the "reality" to be understood, both now and at any time; and, should there be any doubt about it, each of us is a living witness to having had a beginning and to this being a real question: What defines the moment of our beginning? Secondly, the existence of a variety of answers to the moment each of us begins does not necessarily suggest that there is no answer as that the question of "when" needs to take account of each discipline's tendency to answer; for, if the object of our inquiry is a real object, then each discipline is a part of the whole response of the human race to this question. On this basis it is clear that truth will not contradict truth; but, according to whether or not it is embryology, philosophy or theology, the emphasis or preoccupation may be different. Thirdly, taking the unitary view that creation is an act of God, however that is to be understood, there is the coherence of these disciplines in view of the fact that each one of them is answering the question according to the truth that has come to be known or revealed.

Embryology: What is one organism?

What makes an object one? Numerically, it can be argued, one is an adjective which describes a single object;

and, therefore, let us consider a single seed. I am thinking, then, of the courgette seed. This seed is a good example of a very visible seedcase; for, slow as it is, as the temperature rises and the plant begins to grow, it pushes down its root into the soil and, last of all, the leaves slide, as it were, out of the seed-case. Thus there is a very visible plant and seed-case; indeed, as the leaves grow, so the seed-case sits on the edge of the leaf like a hat. In other words, the seed-case is clearly very different to what grew out of it. Now it is not possible to speak of this being like the mother's womb because the womb is integral to the mother; however, just as the plant leaves the seedcase, so the child is born from the mother. The seedcase does not appear to have been alive at all except when it originally grew and, as such, was not alive of itself but even then was a kind of protective sheath of what was dormant and, in due course, ready to grow. Thus the drying out of the seed case is not necessarily the drying out of the seed. Nevertheless, come germination, the courgette seedcase is simply discarded. One seed, then, is both case *and* embryonic plant; and, as the embryonic plant grows so it is clearly separate to the seedcase. In this instance – one seedcase encases one embryonic plant.

Similarly, a child in the womb is in an embryonic sac, both breaking apart the amniotic fluid the child floats in but, at the same time, this cover to shed once the walls breaks and the child is born.

both

One plant, then, is the root and shoot; and, indeed, when I have transplanted a courgette plant and accidentally tore off a part of the pale root, with the little hairs, I have wondered what would happen. It was a young plant with three leaves and two of them shrivelled up and died; but, with watering, the remaining leaf remained healthy and a tiny shoot emerged from a kind of crevice at the top of the plant's stem. Thus, whatever happened underground, the plant regained its more mature root and stimulated the continuing growth of the leaves. A whole plant includes, then, whatever is integral to what makes it one plant: root and shoot. When it comes to a bunch of leek seeds which have been planted in a small pot, they eventually out-grow their container. Shaking the soil off the leek plants shows that their roots are a bit tangled together; however, as they are gently pulled apart it is quite clear that they are a group of separate plants, each of which has now been planted and is growing independently of the others. Just like children conceived from separate eggs, the twins or triplets develop independently of each other as indeed the birth of a girl and a boy, or two girls and a boy make clear.

In the case of runner beans, however, there are many wandering upward growing stems, on which bud many

leaves and from which the red flowers come, prior to producing the runner bean. Thus one plant, in the case of the
runner bean, has a number of variously growing parts all
of which originate from the one seed-case; however, the
seed-case of the runner bean does not look like the seed-
case of the courgette – rather it almost sits above the
ground, having changed its colour or lost its outer lining,
and remains there until the plant uses it up. The potato
plant, though, is a marvellous example of what starts as
one and can be made to twin or even to grow multiple
plants. To begin with there is a single potato, bound by a
rough skin which, often as not, is peeled – especially when
it comes to cooking potatoes that are going to be mashed;
but, before peeling, as you examine the potato there are
growth points called "eyes", which are little indents with
what appears to be a dark centre. "Eyes" are different to
the marks and scratches and holes that cover the surface
of the potato. Often, after a while in a darkish warm cupboard, the "eyes" start to shoot; however, there is one
place where, while there is an indent, there is a kind of
hairy remnant of something – this is the remains of the
umbilical cord from which the potato grew and through
which it was fed, like a narrow tube inflating a balloon.
The "eye shoots", if you look at them closely and even

through an ordinary magnifying glass, look like fresh pale bubbles of potato growing out of the slightest "spot" on the potato and indeed often from many different places on the same potato. These shoots are brittle and will break off if brushed; but, left to grow, there will emerge both plant roots and shoot.

What is amazing, however, is the number of potato plants that will grow from a single potato. As each "eye-shoot" develops, so there is the possibility of having as many plants. Thus, if there are four shooting eyes it is possible that there will be four potato plants. In other words, cutting the potato up so that each shoot has a little of the potato itself, results in as many plants as there are "shooting-eyes" and a bit of potato itself. What is amazing, however, is that a potato plant will grow from as little as a potato peeling – provided that the peeling has an "eye" on it. In other words, there is nothing difficult to follow here: as many as there are potato shoots, eyes and a little bit of potato, there will be as many plants. Thus it is not as if the one potato has ceased to exist; rather, each part of the potato capable of developing into an independent plant has done so.

All in all a human embryo is one human person and, if there is twinning or triplets, then there are two and then three children from the same or multiple eggs. In other words, a little examination of reality shows that there is no confusion about the fact that one divided by another is two: one plus the original one; and, were there to be any possible doubt, then conjoined twins show us the reality of two people from a single, albeit incompletely divided, human embryo. Conversely, if the conjoined twins have arisen from two distinct embryos, known as non-identical twins, partially recombining, then the point remains that two bodies, however combined, are the bodies of two people. In other words, even if at times it is difficult to determine the precise beginning or end of the body of one person and the body of another, it is nevertheless clear that a body expresses the temporal reality of the person in an unrepeatable way: one body expresses one soul and the whole is the human person.

Reverting to what is original

It is clear, in the case of the seed, that the whole plant emerges by degrees from the shedding of the seedcase; and, indeed, when a shoot, particularly the runner bean shoots, are ravaged by slugs and snails, they survive in an albeit slighter form. In the case of the latter, putting a plastic bottle over them or an empty upturned glass bottle helps to protect the struggling plant; and, in its own way, this eating down of the new plant to a stalk argues for putting them out more fully grown and in fact growing the plants in more sheltered conditions. The wisdom, we might say, that would have been passed down were I raised more with a background in growing vegetables. But, as you can imagine, so is it true of the womb, a place within which the newly conceived has the possibility of the nourishment and growth; indeed, just as we discover that a watered-down solution of horse manure is so much more effective than just plain watering, so the richness of the mother's diet enriches the blood supply to the baby. But, more fundamentally, there is presence of the whole interrelationship of mother and baby and the proximity of the father's voice and help. In other words, where there is fertilization in a glass dish, exploitative experimentation of the human embryo, freezing or culling there is clearly a complete disregard for the benefit of the psychological presence of the mother and father, as well as the implicit quality control mentality in judging who will live and who will die: there is a complete disregard of the fact that the child is an unmerited gift and has a right to the completing human development provided in the womb.

Without, however, insisting on a perfection of husband
and wife, there are nevertheless degrees of imperfection
which are dangerous to the welfare of mother and child
and indeed to the father too; and, therefore, just as it is
possible to protect or help plants in view of their natural
predators, either by temporary shelter or even transplanting them, so it is necessary, sometimes, to protect the vulnerable in a specific family.   / *Remind Lr fe*

Returning more explicitly to the theme of unity there
is, surprisingly, what happens to the heart of a cabbage.
Take a cabbage, one already cut from the ground and
packaged, put in a supermarket and taken home, eventually, sliced up and prepared for dinner – and you take the
heart of that cabbage, normally bitter and hard and discarded and you plant it! What do you expect to happen?
That the cabbage starts to discolour and to rot, maybe it
will get a bit eaten as a meal for slugs and snails or simply,
unobtrusively, shrivel away. But would you expect it to
grow into a cabbage plant? But that is what happened –
not once but to three or four cabbage hearts planted in the
soil with a bit of compost and watered! In other words,
and this is the point, there is not only a residual life, calling
into question the actual "death" of the plant but there is a
definite cell development which is not in the least random

but is in fact a reversion to the whole process natural to becoming a cabbage plant – not growing cabbage-like but actually growing into the cabbage plant from which flowers and seeds come!

There is, then, a "residual life" in what appears to be a "dead" cabbage: decapitated, rootless, left on a shelf and cut up for cooking; but, in and amongst the cells of that discarded, bitter heart, are what are still capable of growing, given the ground into which it is planted and the normal conditions of growth. We notice this too with transplanting lettuce, leeks or other plants, that if they are forgotten and left out overnight on the ground they do not just wither and die in a moment but they remain alive and ready to grow for a definite period of time Thus it is possible to speak of a real presence of life that persists even when the conditions for its expression are temporarily withdrawn or problematic. Similarly, in the case of those of us who have died through heart failure or for whatever reason, there is a possibility of being revived; and, as we know with resuscitation, the person can be revived and live and even live a long time after being revived. Thus, the whole concept of dying needs to take account of the residual nature of life and not anticipate death impatiently or even for reasons of organ stripping[15]. Furthermore, the conception of any child, under whatever circumstances, is going to entail a life that persists until it is definitively dead; and, therefore, life is integral to what is conceived and begins to unfold and manifest its personal nature in the course of his or her development. What is more, the life that persists is the life of the whole person and not just some functioning of a part of the whole; and, therefore, where there is the life of a person there is the whole life of that person.

---

[15] The concept of "brain death" has this disadvantage that if the major organs are functioning, such as the heart and lungs, there is the presence of life and therefore of the human life of the person. In other words, the person is not dead.

PART II

LITERARY TRUTHS
AND THE LITERAL TRUTH
COMPRISING CHAPTERS THREE AND FOUR

What is an analogy? An analogy is a partial likeness between two different things in order to help us to understand what is less visible but still true. The point of this brief discussion is simply to indicate how an author can use imagery to indicate partially hidden mysteries; but, in this use of imagery, there is a distinct sense of the author drawing on what exists but not being restricted to a literal use of it.

When God made man and woman in the first book of the Hebrew and Christian Bible, we can see that the opening chapter gives us a varied account of how He brought to exist the whole of creation; but, in that first chapter, the author does not explain how God created man, male and female, although we can see that man and woman are made for "relationship" just as God is a mystery of relationship: 'Let us make man in our image, after our likeness …. So God created man in his own image, in the image of God he created him; male and female he created them'

(Gn 1: 26-27)[16]. In the second chapter, then, the author says: 'the Lord God formed man of dust from the ground, and breathed into his nostrils the breath of life; and man became a living being' (Gn 2: 7). Thus, we see that the creation of man is distinguished from the creation of the animals in that there is a personal action of God which brings the man to exist – both from the earth and because of the breath of God. In other words, God makes man a unified being of both matter and spirit. The emphasis, then, in the making of woman from the man, is that she is "life from life" and, therefore, a sign of "relationship": the relationships hidden in God but indicated in the analogy of man and woman's being made in a unique way. Just as the Son of God comes forth from the Father and the Spirit proceeds from both of Them, so man is made from the earth and the woman is "from" and "through" the man.

In the case, then, of the breath of God, we do not understand breath as indicating the bodily presence of God

---

[16] For a discussion of the three names of God in the first two chapters of Genesis in the light of the Hebrew text, go to Chapter 8 of Francis Etheredge's *Scripture: A Unique Word*:

https://www.cambridgescholars.com/product/978-1-4438-6044-4.

– but the power of God to give, personally, a personal life to man: a life that is intimately and mysteriously shared between the first man and the first woman – each being equally a person. Similarly, when it comes to thinking of the 'dust' from which man is made, it is clear that on the one hand man is made of materials common to the whole of creation – but which are lifeless without the gift of life given by God; but, on the other hand, it is also clear that when God 'formed man' he made this dust capable of bearing the spirit that gave life to him. As another author says later, when it is better understood that there is both a beginning and a development of that beginning: 'Thy eyes beheld my unformed substance' (Psalm 139: 16) and 'For thou didst form my inward parts' (Ps 139: 13) – as the author moves from what is developing to that from which "I" developed; literally, in the Hebrew, "*golmi*", my unfinished vessel![17]

In contrast, then, to this literary use of imagery to communicate a variety of truths, is what constitutes the literal use of understanding one thing, not as like another, but as some kind of real version of the other but in a very

---

[17] Cf. Chapter 11 of Etheredge, *Scripture: A Unique Word,* 2014.

different context. In what follows we will examine a line of thought which took the view of a three-stage stepped beginning to human personhood, the first step of which proposed the existence, literally, of a plant type of being to the beginning of human life.

CHAPTER THREE
PASSING THROUGH THE PAST TO THE PRESENT

The idea that there is not an immediate beginning to being a human person but that there is a process of becoming human has both a very ancient lineage and very modern currency. Therefore it is necessary to examine this kind of account and to show the difference between an author literally understanding the beginning of human being to be that of a plant, or at least not fully human, but becoming so in stages.

From the "literal" use of an image to the truth of embryology

Going back, then, to our discussion on analogy, it is helpful because, as we have seen, there is a partial likeness between two different subjects, the more familiar one helping us to understand the less familiar one; for example, the lifeless 'dust' being formed in such a way that it can receive the breath of life, where breath is commonly understood as a sign of being alive. The contrast of being alive with dust is more about the wholeness of life coming from God than, literally, we are made from the dust of the

ground; however, as a figure of what we have in common
with the matter of the universe, "dust" serves very well.
We now come, then, to a literal use of a comparison between plant and human life. But first a little context to
help us to understand where our philosopher is coming
from.

Aristotle (384-322, BC) was an early philosopher who
researched and wrote on a wide variety of subjects. In general, he held that God was a "First Mover" of what exists
and, unlike the biblical Creator, the "First Mover" did not
bring matter to exist but, rather, brought about changes in
what did from all eternity already exist. Aristotle held that
matter had always existed and so what changed matter,
was the "form" that it received; and, therefore, the "First
Mover" was the origin of all change in matter. On the one
hand, a piece of wood exists but according to what the carpenter does, that wood is a chair leg, a table or a shelf. In
other words, there is what exists and then there is what
changes what exists from one definite object to another –
from a piece of wood from a tree to a chair leg. In the case,
then, of a plant there is that which gives life to inanimate
matter, matter that is non-living, and that is called, in this
case, the plant soul: that which makes matter not just to
be a particular plant but to be a living plant. Similarly, an

animal soul makes non-living matter to be a living animal, which moves, grows and reproduces. Lastly, a human or rational soul makes it possible for non-living matter to be a living human being. On the other hand, to understand the relationship between "matter" and "form" it is necessary to think of "form" as radically changing matter – not just shaping the wood as a carpenter would but bringing about the very change in matter that makes it wood in the first place, or a plant, or an animal or a human being. In other words, "form" intimately determines what matter is: a plant form or soul makes the matter to be a living plant.

Stage One: Taking the comparison with a plant to be literally true

Aristotle, then, considered the first instance of human development to be more than comparable to the growth of a plant. He thought this, mistakenly as we now know, because he thought that the initial condition of human growth was a kind of thickening of blood; indeed, that the male semen brought about a change in the blood which completed the first stage of development. In other words, given the simplicity of what he understood to be the starting point of embryonic development, he could not envisage a more developed or complicated being arising out of the clotting and forming, as it were, of blood into a first being. Thus, his whole understanding of there being a plant type of initial human development is predicated on his belief that little more than liquid ingredients were needed to begin human development. Now, then, it is perfectly plausible that the sperm introduced a kind of "order" or "form" to this "matter" such that it transformed it into a plant type of existence. For, on the basis of blood being the first "matter" with which human being begins, it was clearly impossible for it to be "instantly" developed into what would be capable of receiving an intelligent, rational soul.

Aristotle thought, then, that the first step of human development is literally like the growth of a plant. Thus, he thought that a plant soul inwardly shaped human matter to be a kind of plant; but, as we now know, Aristotle's understanding of a human being's beginning as a plant, actually works very well as an analogy but is no longer literally true. Although the human embryo is like a germinated seed, taking in nutrients, im-*planting* in his or her mother's womb and, through multiple cell-divisions, increases in size and complexity with a more and more recognizable differentiation of parts – it is the growth of what is from the beginning a human embryo and not a plant. In other words, while Aristotle could not appreciate how wonderfully and intricately ordered the human embryo is, and therefore he resorted to a simple explanation of its development, the sophistication of the human egg is such that fertilization initiates immediate and ongoing changes.

Simultaneously, as we now know, some cells develop into the placenta, the organ between the mother and the baby, which moderates and facilitates the transfer of nutrients to the baby and the excretion of waste products from the baby through the umbilical cord to the mother's blood supply and, ultimately, to be excreted by the mother. What has been understood, then, is that the baby begins his or her life as, literally, being planted in the womb of the mother. This is not a diminution of the dignity of motherhood; it is, rather, a true recognition of the fact that the child is not a part of the mother but the mother is "bearing" the child. In other words, the ancient philosopher understood very well that the child has a "beginning" and that that beginning entails concrete steps which manifest and unfold the reality of his or her independent existence.

Stage two: movement, sensation and the rearing of young

The philosopher's second stage of human development is when there is movement and sensation and, while a lot later, the capacity to bear and rear young.

The philosopher, to explain the change from the planting of a human embryo or germinating seed to being capable of movement, suggests that just as there was a plant soul that changed matter into a living plant so there is an animal soul that changes matter into a living animal. We must remember, too, that the word animal comes from an expression "to animate": to make to move. In other words the philosopher was suggesting that just as human development goes through another stage of development, which now entails being no longer planted in the womb but being able to move in the womb, then he or she is given another kind of soul: an animal soul which makes movement possible.

By way of a little explanation, we can see that a plant responds to temperature and that when it is cold a flower will often close up and that when it is warm that same flower will open up; and, more generally a plant, without getting up and walking, will nevertheless lean in the direction of the sunlight and turn, slightly, to follow it. In other words, as simple as it is, the plant is capable of limited movement and thus is ready to receive an animal type of soul that provides, therefore, for the further development of movement. An animal type of soul, enabling movement, makes possible all the characteristics that belong to an animal, such as sensation, digestion, the power to be able to reproduce and the limited horizon of understanding that belongs to making rudimentary dwellings, finding food and rearing young.

In reality, however, we now know that after the implantation of the embryonic child, the developing embryo is rapidly generating the beginnings of its mature form of head, brain, spinal cord, heart and heartbeat, limbs and the whole interiority of organs and their coordinating functions. In other words, movement and sensation are not because of the imposition of a second form on "matter" but because the human embryo, already moving through a tremendous amount of cellular development, is developing according to the inbuilt human pattern, developing all that is entailed in having limbs and, what is required to move those limbs.

Furthermore, too abstract an account of the developing child, or too great an emphasis on intellectual powers, overlooks almost entirely that conception is about relationship; and, therefore, the more "visible" becomes the presence of the child the more explicit becomes the relationship. Loss, therefore, as in a miscarriage, is clearly a fruit of the relationship that has become more and more explicit; relationship being there from the beginning in the very nature of conception, so that even the early loss of the child is expressed, therefore, in terms of the grief of the mother, father and other relatives.

Stage three: rational ensoulment

We are now, therefore, in a better position to understand the ancient philosophical idea of rational ensoulment: that Aristotle, understandably, thought that there needs to be a sufficient degree of development so that the creature could be ensouled as a thinking human being; indeed, a rational human being is more than just a thinking type of human being, for reason structures the whole being in such a way that we are now talking about the whole human being. Thus, as the child matures and readies for birth, there begin to be the signs of distress, contentment, and the rudiments of play.

By now, however, what we are actually witnessing is the interior structure of the human person maturing, physiologically and, therefore, psychologically, so that there is not an additional action of a different kind of soul so much as what is there is increasingly manifest in all his or her rudimentary humanity. By comparison, then, a computer program may be layered in such a way that to begin with there are simple operations that, once mastered, lead to the unlocking of more and more sophisticated capabilities. In other words, it is not that there is present a new kind of soul as that physiological development allows the showing forth of the more subtle, expressive and intellectual potentialities which are now able to start to show themselves. Indeed, just as a child masters walking and running, so he or she begins to demonstrate the existence of balance, control of a ball or even the dexterity that is a part of being a musician.

While we have seen that this ancient idea that the human person began in stages; and, while it had some credibility in the past, what we now see is that it was a "theory of gaps": an attempt to explain human development through a comparison with a three-step ladder of being. The problem that this ancient idea attempted to solve was how to go from a simple beginning to a stage of development that would make it possible for God to bring the whole human person to exist, one in soul and body. In other words, it was thought that the very beginning of embryological development was too simple to be the moment in which God could create the human soul.

A concluding reflection – towards understanding human ensoulment

Using what was perhaps more literally understood by the ancient philosopher now, as an analogy, helps us to see how to understand a whole plant, the independent conception and growth of the child and the confirmation, of birth itself, that a child is no more a "a part" of the woman's body than a plant is a part of the soil in the ground. In other words, just as a plant in the soil can germinate, grow, be transplanted and ultimately uprooted if necessary, so it is obvious that the child is likewise "implanted" in the womb of the mother, develops and is delivered at birth. Just as the unfertilized egg is from the mother and constitutes, in its own way, the basis of the bodily being of the human embryo on being impregnated by the father's fertilizing sperm, so the lining of the womb is so ordered to the implanting of the embryo and, therefore, the dialogue of growth, such that the developing embryo can draw what he or she needs to grow from the mother. Naturally, then, the diet and health of the mother are equivalent to providing an enriched soil for healthy plants. What is more, however, the presence of the developing embryo in the mother is also expressive of the centrality of "relationship" to the formation of the humanity of the human being.

In contrast, however, to the progressive replacement of a plant, animal and rational soul, it is now abundantly clear that there is a seamless development from conception to the full manifestation of the presence of the person. Furthermore, what is clearer to us than to ancient philosophers is that the embryonic human child is far more developed than was hitherto understood and, therefore, given that the human soul is the life of the body, no sooner does the body come to exist than the ensouled, whole person, one in body and soul, is present. In other words, although we cannot see the soul, philosophically, if there is no soul then there is no human body – for a body is not a body if it is not specifically animated by a soul just as there is no such thing as a husband or wife who is not a particular person. This is because the unity of body and soul is so intimate that there is no "generalized body" but only the specific human life, bodily expressed, which is ensouled and shows itself to be personal from the first instant of conception, expressing as it does a unique but totally human genetic inheritance.

The greatest natural transformation: the unfolding of conception

Modern human embryology, then, is far more definite about what exists from the first instant of human fertilization: from the beginning[18]. Indeed, right from the first instant of a fertilizing sperm in an open pore of the egg, the calcium ions rush from the sperm and across the newly formed human embryo, closing the embryonic wall. Indeed, everything has a precise "moment" and significance in terms of timing, direction, sequence and function – so that the whole being of the human person undergoes what is probably the greatest natural transformation in nature: the unfolding outward expression of the inward presence of the person from conception. Just as

---

[18] Cf. For an excellent introduction to human embryology, go to: Chapter 5: Part II of Etheredge, *Conception: An Icon of the Beginning*: https://enroutebooksandmedia.com/ conception/, where you will find an account of early embryology by Professors Justo Aznar and Julio Tudela.

whatever is biologically alive, expresses its life electrically, so a current being on is inseparable from what transmits it; indeed, just as yeast introduced into dough, spreads and makes it rise, and is no longer distinguishable from it, so human life once begun is irreversibly human. Conversely, it is clear that the various disabilities are not about being anything less than human but about the challenge of communicating that humanity when its natural expression is impeded, frustrated or, in some measure, is dependent on an adequately helpful technology.

Note that the transformation entailed in a human person coming to exist unfolds an inner nature scarcely comprehensible in its initial and first formed stage of being a human embryonic person until, literally, there is the manifestation of the person present from conception. In other words, what if we could run the development of a beautiful flower backwards until it was the first instant of a germinating seed, would it help us to appreciate that running the development of a human person backwards takes us to the first instant of human conception?

Why cannot a human soul be transmitted through the process of human procreation? The human soul is expressed in the body as a whole; indeed, the soul is the form of the body: the soul is that which determines the body to be Paul or Janet's body. Given the life of a plant and an animal, what they pass on is within the power of each to reproduce: to literally replicate what it is to exist as a plant or an animal. However, in the case of a human being, there is an element of organic life which is transmitted through the activity of the sperm – but the power of human thought transcends the power of an organic transmission of life; and, therefore, there is an act of God which establishes the whole human being at conception – not just as a general kind of human being but the human being of a particular human person, capable of expressing what is beyond the capacity of either plant or animal life to express, namely truth and love.

While, then, there are many ways of putting our comparisons between the beginning of one thing and that of another, there is also the literal truth that there is a beginning to human life to which each one of us is a concrete witness.

CHAPTER FOUR
SCRIPTURE AND THEOLOGY: WORD AND DOGMA

The progress of science, however, has shown that from the first instant of contact between sperm and egg, they form the new entity of an enclosed, walled and dynamically developing human embryo. Thus, it is now clear that the dynamism that begins with the sperm's entering the egg, bringing about an irreversible change that makes present the beginning of embryonic development, is a dynamism that unfolds the uninterrupted presence of the person from conception. Indeed, as the embryological development unfolds, so we see that the child's physiological and psychological development shows itself in his or her movements and responses to both internal changes and to the stimuli of being-in-relationship to others – particularly the mother, the father and the immediate family and friends who are "with" and look forward to meeting the child. The psychological development, embedded and expressed in the unfolding of the anatomical and physiological developments, shows that there is present, from the first instant of fertilization, an identity which goes beyond what biological matter, whether sperm or egg is capable of. This is particularly clear in that the uncombined sperm and egg deteriorate, showing that they are transmitting what is biologically required for human life but which are an insufficient explanation of the fullness of human personhood.

The activities of the child which show forth his or her humanity are varied, ranging from responsiveness to the mother and father, humour, expressing the psychological states indicating hunger and thirst and, little by little, taking up the common inheritance of the language and beginning to express ideas. As the emergence of language acquisition, thought and humour show, the child expresses a being which goes beyond the capabilities of organic matter. Thus, the psychologically personal life of the child expresses the presence of what goes beyond plant and animal life, namely the presence of a human soul; and, in view of the seamless, uninterrupted nature of human development, the moment of the existence of the embodied soul is therefore the first instant of fertilization. What causes the existence of the embodied soul, and therefore that of the whole human being, has to be a cause capable of bringing to exist what did not; and, therefore, just as the powers of the human embryo show the presence of what exceeds the capabilities of matter, so the cause of human personhood exceeds that entailed in what the sperm and the egg can contribute to human existence. What the whole of human identity and development communicates to reason, then, is the presence of a cause that exceeds human existence, namely God, just as human existence exceeds the limits of the organic matter through which it is expressed; and, therefore, the existence of the person reveals an action of God which brings about a uniquely embodied soul from the first instant of human fertilization.

The unfolding of that first instant is one of the most amazing journeys of being: that the inwardly present soul, because of the creating cause of God, determines the outward expression of the whole human person. The scientific evidence is overwhelmingly clear that from the first instant of fertilization there is an autonomously functioning being that is wholly independent of the parents and, at the same time, is in relation to both: to both the father and the mother. Each one of us is, then, an implicit witness to the beginning of a new being: the specific person that each one of us is.

The mother, being the bearer of the child, expresses more completely the origin of the child in the love between the father and the mother; and, in addition, the essential truth that relationship is at the core of human existence and development. Thus, the mother's welcoming receptivity is both an expression of what she contributes and is its own evidence of relationship being the fundamental vehicle of human development. This truth of science, both embryological and psychological, is confirmed abundantly from the very different but complementary account to which we now turn.

In the Scriptural text we will see that there is a very personalistic account of the conception of the whole human person, implicitly understood as a "living being"; and, as such, each one of us is an intimately personal work of an intimately personal God. In other words, we see that the word of God completes, in a way, our understanding of human conception and development by taking us further and further into the mystery that *relationship* is the great and understated mystery evident, but not explained, throughout the natural world and particularly present in the conception and development of human life.

But first we must consider, however briefly, on what basis we can justify our use of the biblical account.

The word of God and dogma

It can seem to us that the word of God, the Scripture, is not a truly human word or not really a truly divinely inspired word. We might, for example, argue how is it possible for a human author to write anything that is absolutely true when, as we all know, human beings are so prone to mistakes or to find some subjects so difficult to understand that they remain almost totally impenetrable: On the other hand, how is God going to be able to communicate through a word which is so obviously written through all the problems and sins of the human race to the point that one wonders is it even possible that the writer can be sufficiently knowledgeable or open to divine inspiration that, notwithstanding all the author's defects, he can still write *the truth which, for our salvation, God has confided to the sacred Scripture (Dei Verbum*[19], *11)*? So we come upon all the challenging questions which surround the use of Scripture in a discussion such as this and, surprisingly, we find an unexpected answer in the dogmas of

---

[19] The documents of the Church are generally known by their first Latin words and the number that follows is the paragraph numbering given by the Church when the document is published; *Dei Verbum* means the word of God, literally, and is a text from the Second Vatican Council explaining the nature of Scripture and its relationship to the teaching office of the Church and to Tradition.

the Church: that there is a certain understanding, made possible through the help of the Holy Spirit (*Dei Verbum*, 5 and 8) which takes us to what is essential for our salvation. 'One … of Cardinal Ratzinger's most famous statements that dogma is simply the Church's infallible interpretation and elucidation of Scripture'[20]. Moreover, we find the help we need in an unexpected dogma concerning Mary, the Mother of the Lord; but, before we consider Mary, the second Eve, we will start with a few other voices, beginning with the first Eve.

A variety of witnesses to human conception – beginning with Eve

'Now Adam knew Eve his wife, and she conceived and bore Cain, saying, "I have gotten a man with the help of the Lord"' (Gn 4: 1). This is the earliest, simplest, truest, comprehensive and most enduring account of human conception that we have; and, indeed, it will take millennia to appreciate and grasp, more and more fully, the

---

[20] Dr. Joshua Madden, "Newman, Aquinas, and the Development of Doctrine": https://www.hprweb.com/2021/06/newman-aquinas-and-the-development-of-doctrine/.

fullness of truth that is expressed by both Eve and the biblical author. On the one hand the biblical author recognizes that 'Adam knew Eve his wife, and she conceived'.
Thus intercourse is an intimate "knowing" between husband and wife and, by implication, involves the wholly
personal communication between them which is, at the
same time, a moment that arises out of the social reality of
their marriage: a wholly reciprocal gift of husband to wife
and wife to husband of all that each of them is. On the
other hand, the author quotes Eve as knowing and saying
"I have gotten a man with the help of the Lord" (Gn 4: 1);
and, therefore, Eve has an equally intimate sense of the
Lord's help giving her 'a man'. Indeed, so significant does
she regard 'the help of the Lord' in conceiving 'a man' that
she does not even consider what the narrator tells us,
namely, being known by Adam; rather, so conscious is she
of the 'help of the Lord' that this is her primary perception: that without the 'help of the Lord' there would be no
conception. In other words there is a dramatic sense of the
act of God which both manifests the continuation of the
Creator's act of creation and its essential contribution in
bringing a human being to exist.

Job

He builds on Eve's perception, making it his own when, complaining about his own life Job says that God has taken a stand against 'the work of thy hands' (Job 10: 3). Elaborating on his beginning, Job goes on to say: thy 'hands fashioned and made me' (10: 8) and, as if recalling the second account of creation, Job asks God to remember, as he himself recalls, 'that thou hast made me of clay' (10: 9). Job then takes a further step and, drawing on his own imaginative analogy, he starts to analyze the "moment of his conception"; he compares his origin with the curdling of cheese: 'Didst thou not pour me out like milk and curdle me like cheese' (cf. Job 10: 10). Thus Job has expressed himself in a new way, going further into the possibility of the action of God in the mystery of human conception. In other words, taking the literal image of being poured out 'like milk', suggesting that God has not only poured him out but that there is also a likeness between being poured out and milk, implying some understanding of what is involved in human conception, possibly of the semen being like milk and that this is a stage in a process, after which God then curdles him 'like cheese'.

Without, then, attempting to identify too closely this analogy between milk and cheese and human conception, yet one can recognize the imaginative connection between what we now know and what, to some extent, Job surmised from what he knew happened. What distinguishes this process of Job's, however, in contrast to the philosophical steps of three types of soul, is that Job is Job from the beginning, through to the end. In other words, there is a sense of personal identity that runs through the whole event from when Job first recognizes that he exists because of the action of God: he is the 'work of thy hands' (Job: 10: 3); and, by way of continuing emphasis, each stage is still a stage of Job being made: 'Thy hands fashioned and made me' (10: 8) right up to the possibility of him going to the 'land of gloom and deep darkness' (10: 21). By contrast, then, the early philosophical account, for all its richness, does not address a wholly human identity until the third step of rational ensoulment (following the plant and animal stages of development).

Moreover if, taking the relationship between man and God from Genesis as that we are made in the image and likeness of God (cf. Gn 1: 26), then it follows that Job is addressing God as a "personal being" the whole time as, from start to finish, he addresses God as the author of

Job's being: Job is the work of 'thy hands' (Job 10: 3) and, when coming to the depth of his suffering he says: 'Why didst thou bring me forth from the womb?' (Job 10: 18).

David

The author of Psalm 139 similarly, recognizes the action of a personal being who has made him: 'For thou didst form my inward parts, thou didst knit me together in my mother's womb' (Ps 139: 13); and, using a unique Hebrew word which occurs only once in the bible, *golmi*, David says, 'Thy eyes beheld my unformed substance' (Ps 139: 16). The expression, 'unformed substance', more literally means 'unfinished vessel' and, as such, implies what has been made, which is a container, containing either a soul or the grace of God or both, that it is unfinished and therefore developing, and that God is making "my" unfinished vessel as David also says: 'for thou didst form my inward parts' (Ps. 139: 13). In addition, however, David also uses the imagery of a book which is, after all, a personal work of an author; and, therefore, both he and Job, while not giving exactly technical account of their conception are nevertheless indicating sufficiently clearly that there is a beginning and a process of development; and, in

their own way, therefore, they are very reasonable accounts of human conception within the limitations of the
times in which they were written and the language that
was available to them. Although, having said that, David's
word, *golmi*, seems to have been specially coined for the
task since, as I have said, there is only one instance of its
use in the whole Hebrew bible; indeed, a unique word for
a unique event of the action of God and the beginning of
a human life.

The Martyred Mother of Seven Martyred Sons

Finally, and somewhat surprisingly, the most explicitly
scientific account of human conception and development
is from a Martyred Mother of several martyred sons; she
says:

'I do not know how you came into being in my womb.
It was not I that gave you life and breath, nor I who set
in order the elements within each of you. Therefore the
Creator of the world, who shaped the beginning of man
and devised the origin of all things, will in his mercy
give life and breath back to you again, since you now

forget yourselves for the sake of his laws' (2 Macc 7: 22-23).

Furthermore, by way of complementing this more speculative account of human conception, referring to 'life and breath' and the fact that the 'elements' were 'set in order ... within each' of her sons, this mother goes on to speak clearly of the process through which one of her sons came to exist: 'I carried you nine months in my womb, and nursed you for three years, and have reared you and brought you up to this point in your life' (2 Macc 7: 27). Calling into the argument the witness of what God made out of 'what did not exist' she went on to say that 'Thus also mankind comes into being' (2 Macc 7: 28); and, therefore, that if God can make all that exists out of nothing so He can return her sons to her: "Accept death, so that in God's mercy I may get you back again with your brothers" (2 Macc 7: 29). This Mother's reflection on the coming to be of her sons is all the more remarkable as she witnesses their martyrdom before, finally, being martyred herself for refusing to reject the binding laws of God. This Mother's testimony is a wonderful combination of faith and reason; faith in the power of the Creator which is, in addition, married to a 'woman's reasoning' (2 Macc 7: 21);

and, in the context of her whole testimony, this 'woman's reasoning' is both scientific and experiential, both observant and penetrating in her analysis.

Mary: The dogma of the Immaculate Conception and human conception

We now come to the final point of this examination of the Scriptural evidence for the beginning of each human life, excluding the creation of Adam and Eve. What we find is that the Church has expressed herself in the dogma of the *Immaculate Conception* of Mary, the Mother of our Lord Jesus Christ. A dogma, of which there are other examples, such as the Assumption of the Blessed Virgin Mary into heaven, one in body and soul, on her death, is an expression of the Church's certainty that the doctrine proclaimed is not just true but of a truth so necessary to the whole of salvation that we are called to recognize it as integral to our salvation; and, therefore, to disbelieve or to reject a dogma seriously imperils the wholeness of our faith in what God has done to bring about our salvation in Jesus Christ.

Thus the *Immaculate Conception* of Mary is that she was conceived full of grace to be a fitting vessel of bearing

the saviour of the world; and, therefore, in view of the
merits of her son Jesus Christ, Mary is conceived as integrally whole, ordered in herself and open to God to whom
she turns and who makes it possible for her to do good, to
love and speak the truth without imperfection. To put the
Immaculate Conception negatively, Mary was conceived
without original sin: without the tendency to sin transmitted to the whole human race through the process of human generation. In other words, Adam and Eve lost the
integral gift of being created in a state of original justice –
both good themselves and in a right relationship to God,
to each other and to creation. Adam and Eve's original sin
of disobeying the commandment of God was not only a
personal sin but also, as our first parents, entailed the loss
of the graced ordering of human nature which they had
received from God. While we are given baptism, a trini-
tarian triple immersion into water and the mystery of the
death and resurrection of Jesus Christ, an act which begins
to heal the wound of original sin and with which we are
called to cooperate, Mary was given a radical immersion
in the redemptive grace of Jesus Christ from the first instant of her conception. Hence Mary was conceived both
full of grace and without the wound of original sin.

With respect, then, to the theme of this final section on dogma and the Scripture, the dogma of the *Immaculate Conception* of Mary entails the view that she was conceived naturally, by her parents. In order for Mary to be conceived without original sin, which is transmitted as a deficiency of the original grace given to Adam and Eve and passed down through human procreation, then it follows that Mary must have been conceived, ensouled and given the grace of being immaculate from the first instant of her conception. If Mary was not given grace from the first instant of her conception then she would not been perfectly free from original sin – because the deprivation of original sin is transmitted through the act of procreation. Therefore, the dogma of the *Immaculate Conception* is a certain interpretation of the moment of human conception being the very first instant that a human being is one in body and soul; and, as such, give a precise meaning to what the author of Genesis and Eve said: 'Now Adam knew Eve his wife, and she conceived and bore Cain, saying, "I have gotten a man with the help of the Lord"' (Gn 4: 1).

The witness of Eve, as of so many others, is not just to a vague sense that God exists and acts; it is, rather, indicative of the whole biblical tradition that God is not defined abstractly – but acts in the concrete moments of existence. While, then, there are no doubt many questions about the beginning of creation and the beginning of each one of us, there is also the simple, literal truth, that where the human person comes to exist so God has acted to complete it.

PART III

WHAT IS CERTAIN AND WHAT IS UNCERTAIN ABOUT CONCEPTION
COMPRISING CHAPTERS FIVE, SIX AND SEVEN

What better starting point, then, than concrete and familiar phenomena and from there to go on to what is more difficult to understand but which belongs, as it were, to the same universe of truths? In other words, what is the meaning of conception, of human conception, if not the beginning of the life of the person; for, by definition of a person coming to exist, the person's coming to exist has a beginning.

Discussing the ordinary propagation of plants helps us to understand in a simple, concrete way, the natural order by and through which one kind of being reproduces another; and, in this simplistic sense, it anchors the imagination that can envisage a whole plethora of possibilities which, however, need a basis in reality if they are to be relevant. In other words, in view of the multitude of possibilities which an imagination can generate, itself going beyond the evidence speculatively and implying the existence of that which enables this to be possible, namely the soul informing the body and constituting the whole of human personhood – there is a need to think in terms of what actually exists.

As we have seen there are characteristic patterns of development where, for example, runner beans develop into runner bean plants and, in due course, produce runner beans for eating or planting. In other words there is a simple reality to there being, ordinarily, kind from kind; and, therefore, this is the case too when it comes to the transmission of human life: it is "life" from "life". Just as a bean is a bean and brings about the existence of plants and beans, so it is human beings that bring about human beings – but in this one respect differently, namely, through the action of God completing the whole by the integrated ensouling of the human soul.

There are, then, a number of certainties and uncertainties: Each one of us is a witness of having come to exist; and, having come to exist, having a beginning. However, there is no certainty, ordinarily, about which act of spousal love will contribute to the transmission of human life and draw from God, as it were, His ensouling action. At the same time as some may argue that we cannot be certain about whether or not we exist, whole and entire, from the very first instant of fertilization or not, we are certain that life is transmitted from the very first instant of fertilization. The sperm, as well as contributing its genetic inheritance and all that arises out of the specific characteristics of coming to rest in the ovum or egg, is at the same time totally energizing and sets in motion irreversible changes that are expressed in the closing of the now embryonic wall and the firing of the mitochondria and the driving of an uninterrupted process of development which, unimpeded, discloses by degrees the presence of the person[21]. The natural moment, then, for an absolute beginning to human life is the very first and irreversible moment of fertilization; and, therefore, this constitutes a nature sacrament: an outward sign of the inward action of God bringing the whole person to exist from the very first moment of fertilization. But, in view of the natural uncertainty as to whether or not a child has come to

---

[21] Clearly, to suppose that there is no clear moment of beginning is more to do with the inadequacy of our perception than that there is not one; and, therefore, just as we take the outward sign of baptism, or any sacrament, as a sign of the inward action of God, so we can take what is clearly an outward sign of the beginning of human life as a definitive sign of the inward action of God – it not being possible to prove otherwise (cf. *Evangelium Vitae*).

exist we are, in fact, called upon to be patient in our perception of the coming to be of a child. In other words, the
natural uncertainty about a child coming to exist expresses, in its own way that each child is a gift; and, therefore, we are all a gift given to give thanks for the gift of life
and to appreciate that we are all equally a gift.

Conversion to reality[22]

Watching plants grow stirs us to think about what is
consistent in nature and, as such, leads us to consider the
significance of this consistency; and, therefore, we can see
that the reality of human beginning provides its own

---

[22] Cf. "Is Faith Married Reason?", Chapter Ten: Part I (of
Volume III-Faith is Married Reason, Newcastle upon Tyne:
Cambridge Scholars Publishing, 2016. This paper, now in-
cluded in the above book, was first presented as a response to
an address by Bishop (now Cardinal) Angelo Scola; the
Bishop's paper was called "The Nuptial Mystery at the Heart of
the Church" (Oxford Catholic Chaplaincy, 21 March, 1998).
The Bishop's paper is cited in this essay as: Angelo Scola, "The
Nuptial Mystery" and page number.

evidence to those who come into contact with the beginning of a human person.

In the following remarkable, and wonderful account of what took over seven years of pondering what was happening in the *in vitro* fertilization industry, we can see an embryologist discover that the reality of human embryonic development is a witness, in its own way, to the presence of the person from the beginning:

'I can remember always instinctively pausing each time my hand was over the biohazard disposal container, as if to double-check that I was, in fact, discarding the correct [human] embryos. Maybe it was also something else that caused me to pause.'

And he goes on to say, especially concerning the uncertainty around what was or was not a "viable" human embryo:

'Despite having some cells that are abnormal, an early embryo has the capacity to "self-correct." It does this by selectively pushing the abnormal cells out and replicating the normal cells. In fact, it appears that the abnormal cells become part of the placenta, leaving the normal cells to become the fetus. This is a remarkable biological process that is somehow "programmed" into these primitive cells.

To me [he says] this revelation reinforces the meaning of the Psalmist's words:

"For you formed my inward parts; you knitted me together in my mother's womb. I praise you, for I am fearfully and wonderfully made. Wonderful are your works; my soul knows it very well. My frame was not hidden from you, when I was being made in secret, intricately woven in the depths of the earth." Psalm 139:13-15'[23].

---

[23] *Dr. Craig Turczynski is a Reproductive Physiologist, Certified Teacher of the Billings Ovulation Method®, Director of Strategy and Scientific Affairs for BOMA-USA, and currently serves on the board of advisors for Sacred Heart Guardians and Shelter:*"Abnormal Embryos and Human Life: An Embryologist's Post-Conversion Reflection": https://sacredheartguardians.org/abnormal-embryos-and-human-life/. Other testimonies can be found elsewhere: Laura Elm, "Foreword to Chapter Three" of *Mary and Bioethics: An Exploration*, by Francis Etheredge: https://enroutebooksandmedia.com/maryandbioethics/; and in *Within Reach of You: A Book of Prose and*

In the end, then, as with any inquiry, the inquirer has to address wholeheartedly and in all honesty the question of what, really, is the obstacle or obstacles to knowing the truth that each one of us has a beginning. May evidence and God Himself, the author of life, inspire whatever progress we may need to make to understand conception further, to recognize it more fully, and to rejoice more humbly in front of the amazing reality that we "behold". Let us remember, too, that the Lord wants no one to be lost; and, therefore, forgiveness awaits us all: He 'is forbearing toward you, not wishing that any should perish, but that all should reach repentance' (2 Peter, 3: 9).

Prayers, by Francis Etheredge, on p. 77, along with footnote: 27: "How many embryos are frozen in China's IVF clinics?"

by Michael Cook, 24 Jan 2021: https://www.bioedge.org/bioethics/how-many-embryos-are-frozen-in-chinas-ivf-clinics/13677. *Within Reach of You* is published by En Route Books and Media:

https://enroutebooksandmedia.com/withinreachofyou/.

CHAPTER FIVE

THE TEACHING OF THE CHURCH AND THE PROBLEM OF UNCERTAINTY

Introduction:

There are numerous directions that the exploration entailed in this book can go. However, in terms of fundamental points, the definition of conception is both required by a thorough understanding of the integrity of human being and, at the same time, extends throughout a whole range of issues: issues that, in the end, have a direct bearing on a person at the beginning of his or her life. The advantage, now, of discussing a specific legal case is that it summarises many of the problems surrounding a clear exposition of the beginning of the human person; but, at the same time, it shows that the implications of the truth unfold, ethically, throughout a whole range of modern developments and grounds, ultimately, in the possibility of a world-wide bio-legal declaration or law pertaining to the human race.

Prologue: A Modern Moment

Providence gives us many gifts and one of them is the help in a particular historical moment to develop our understanding of a truth that is related to our faith – even to a dogma of our faith: the moment of human conception; indeed, as the document on the Word of God, *Dei Verbum*, says:

'This tradition which comes from the Apostles develops in the Church with the help of the Holy Spirit[24]. For there is a growth in the understanding of the realities and the words which have been handed down. This happens through the contemplation and study made by believers, who treasure these things in their hearts (see Luke, 2:19, 51) through a penetrating understanding of the spiritual realities which they experience, and

---

[24] Footnote 5 in the document is: cf. First Vatican Council, Dogmatic Constitution on the Catholic Faith, Chap. 4, "On Faith and Reason:" Denzinger 1800 (3020); from:

https://www.vatican.va/archive/hist_councils/ii_vatican_council/documents/vat-ii_const_19651118_dei-verbum_en.html.

through the preaching of those who have received through Episcopal succession the sure gift of truth. For as the centuries succeed one another, the Church constantly moves forward toward the fullness of divine truth until the words of God reach their complete fulfilment in her' (8).

In other words, the reality of participation in the development of our faith is multi-personal and multi-disciplinary; and, in addition, there is progress in the truth which brings faith and reason into a renewed dialogue: a dialogue that has, in some way, to go beyond recriminations, regrets and even accusations[25]. For, bearing in mind the enormous implications of how the world has lived in the light of the uncertainty about the beginning of human life, we need to proceed with an amazing interrelationship of truth and love. But, at the same time, we are on the threshold of untold developments in which what was thought to be an ethical development is in fact in danger of being the

---

[25] Cf. Pope Francis, *Fratelli Tutti*, articles 226-227 : https://www.vatican.va/archive/hist_councils/ii_vatican_council/documents/vat-ii_const_19651118_dei_verbum_en.html.

total opposite. Thus there are frozen human embryos which have not only suffered the real frustration of the right to natural, completing human development, but which are now being proposed for some kind of mass production whereby they have been 'prevented from reaching their full potential by depriving them of the extra-embryonic cells required for implantation into the uterus, [whereas] … if they, along with the extraembryonic cells, were implanted into a surrogate uterus, [they] could develop into a living human infant'[26]. Therefore, there is an urgency to clarify what we now understand to be the moment of human conception: a conception which establishes the human relationships regulated by human rights; for, not only are we conceived-in-relationship but we are conceived in a relationship of reciprocal rights from conception onwards.

---

[26] "Mass Production of Human "Embryoid" Cells from Developmentally Frozen Embryos: Is it Ethical?":

http://www.cmq.org.uk/CMQ/2020/Aug/mass_production_of_embryoid_cell.html.

Introduction: Who is My Neighbor?

The *Congregation for the Doctrine of the Faith* said: 'Beginning at conception, children suffering from malformation or other pathologies are *little patients* whom medicine today can always assist and accompany in a manner respectful of life. Their life is sacred, unique, unrepeatable, and inviolable, exactly like that of every adult person' (the *Good Samaritan, Samaritanus Bonus*, 6[27]).

Why, however, begin a discussion on the nature of conception with a reference to a document which, generally, attends to the sensitive care and treatment, as well as the accompaniment of, people approaching the end of life and the help that they need: particularly the help that they need to hope in God and to experience the love of neighbour, whether that neighbour is a member of his or her family, a doctor, a nurse or any person with whom they come into contact? Indeed, 'neighbour' is a word rich in

---

[27] Congregation for the Doctrine of the Faith:
https://press.vatican.va/content/salastampa/en/bol-lettino/pubblico/2020/09/22/200922a.html.

significance[28]: 'neighbor (n.): "one who lives near an-other," Middle English neighebor, from Old English ... "one who dwells nearby," from neah "near" ... "dweller," related to bur "dwelling," from Proto-Germanic * ... "to be, exist, grow".' In other words, there are many who live near another, dwell nearby and are, whether literally near or close to the heart of others in need, a neighbor: a con-crete help to others. Who, then, can be a closer neighbor to the unborn than God Himself; and, therefore, who bet-ter to communicate the truth concerning conception than the Church? As *Gaudium et Spes* says: 'For by His incar-nation the Son of God has united Himself in some fashion with every man' (22[29]); and, therefore, if this is a union prior to baptism what moment is prior to that of concep-tion and corresponding, as it were, to the moment of the *Incarnation*?

---

[28] The definition of neighbour was provided by Mr. Martin Higgins, MA, an Eastern European Linguist.

[29]   http://www.vatican.va/archive/hist_councils/ii_vati-can_council/documents/vat-ii_const_19651207_gaudium-et-spes_en.html.

The Problem of Uncertainty in Both Church Teaching and the "Opinion of the Court" in Roe v Wade

But to go back to the question: Why, however, begin a discussion on conception with reference to a document on hope and compassion for the vulnerable, whether elderly or just conceived? There are two, if not three reasons for beginning in this way. Firstly, the compassion and hope which fills this document is as applicable, as it suggests, to the needs of the unborn, their mothers and fathers, as to the needs of the vulnerable at whatever time and stage of life; indeed, especially where 'Beginning at conception, children suffering from malformation or other pathologies are *little patients* whom medicine today can always assist and accompany in a manner respectful of life' (*Samaritanus Bonus*, 6). Secondly, yet again the Church has used the expression, concerning the beginning of human life: 'Beginning at conception, children ...' (*Samaritanus Bonus*, 6); and, therefore, the Church is clearly articulating a constant teaching that a child is conceived at conception. In the *Gospel of Life*, Pope St. John Paul II says: Every person can come to recognize 'the sacred value of human life from its very beginning until its end, and can affirm the right of every human being to have this primary good respected to the highest degree' (2^{30}); and, indeed, what does conception ordinarily mean but the 'very beginning'? *Donum Vitae*, on the *Gift of life*, says in its introductory comments: that *'the first part will have as its subject respect for the human being from the first moment of his or her existence'*[31]; and, therefore, can he or she exist from the first moment of his or her existence without being a person? It is certainly true that a person can exist without being recognized to be present; and, as such, there is a process through which that presence, which is ordinarily hidden, is made increasingly visible and recognizable. Thus, as even the Warnock Report admitted, there is ordinarily a seamless unfolding of each one of us from conception onwards. Thirdly, again the Church speaks of the child's life being 'sacred, unique, unrepeatable, and inviolable, exactly like that of every adult

---

[30] *Evangelium Vitae*: http://www.vatican.va/content/john-paul-ii/en/encyclicals/documents/hf_jp-ii_enc_25031995_evangelium-vitae.html.

[31] *Donum Vitae*:

https://www.vatican.va/roman_curia/congregations/cfaith/documents/rc_con_cfaith_doc_19870222_respect-for-human-life_en.html.

person' (*Samaritanus Bonus*, 6); and, as such, echoes Pope Paul VI and Pope John XXIII, who said: "'Human life is sacred—all men must recognize that fact," Our predecessor Pope John XXIII recalled. "From its very inception it reveals the creating hand of God"[32]'.

The Problem of Uncertainty in both Church Teaching and the 14th Amendment

On the one hand in 1995, in *Evangelium Vitae, The Gospel of Life*, Pope St. John Paul II quotes from the Declaration on Procured Abortion, published in November, 1974, which said: 'modern genetic science ... has demonstrated that from the first instant there is established the programme of what this living being will be: a person, this individual person with his characteristic aspects already well determined' (60)[33]. The phrase which attracts

---

[32] *Humanae Vitae*, 13: http://www.vatican.va/content/paul-vi/en/encyclicals/documents/hf_p-vi_enc_25071968_humanae-vitae.html.

[33] *Evangelium Vitae*: http://www.vatican.va/content/john-paul-ii/en/encyclicals/documents/hf_jp-ii_enc_25031995_evangelium-vitae.html.

attention, in this context, is where the declaration speaks of what has been established: 'modern genetic science ... has ... established ... what this living being will be: a person'. In other words, 'modern genetic science' has established the possibility that there 'will be' a person; and, if there 'will be' a person, then when will that person be there? When will the person who will come to exist – come to exist from the existence of 'the living being' which does exist? Thus there seems to be a distinction between the 'living being' which has come to exist and the person which will come to exist. What has come to exist, then, that is a 'living being' and yet is not a person? As a kind of explanation of the possible thinking behind this distinction between a 'living being' and the person that that 'living being will be' there is a note that was added to the English translation of *Donum Vitae, The Gift of Life.*

'The point of this part of the discussion is to show that, in fact, there was a dimension of meaning that "seemed to be omitted by addition" in the *English* translation of *Donum Vitae,* namely, that the zygote comes to exist 'when the nuclei of the two gametes have fused'. Thus the additional English expression, 'the nuclei of', is both an additional phrase to what is in the Latin text and, at the same time, it is a phrase that seems to de-limit the definition of a zygote to the fusion of the two nuclei. However, as we shall see, the Latin expression, '*orta a fusione*' (arising from a fusion), does not men-tion nuclei and, therefore, is a more comprehensive ac-count of the nature of the zygote. Thus the Latin text may include, in its range of meaning, the development of the zygote from the first instant of fertilization; and, therefore, the Latin may well be more inclusive of that first instant than the English expression: 'when the nu-clei of the two gametes have fused'. The expression, 'when the nuclei of the two gametes have formed' tends to make one think that the zygote has not formed until the nuclei have fused; and, as such, could be described as an interpretative translation, referring to a develop-mental point which is not so clearly evident from the Latin text. For there is a case to advance that the Latin expression, '*orta a fusione*' (arising from a fusion), could apply from the very first moment that fusion oc-curs: the first instant of fertilisation'[34].

----

[34] Excerpt from Francis Etheredge, *The Human Person: A Bioethical Word*, p. 369, St. Louis: En Route Books and Media, 2017.

On the other hand, on January 22, 1973, the U.S Supreme Court legalized abortion in Roe v. Wade; and, without going into the details of the whole judgement, pronounced that the 'State … has legitimate interests in protecting both the pregnant woman's health and the potentiality of human life'[35]; however, the wording of the "Opinion of the Court" is inconsistent: it both refers to the 'potential life' and 'fetal life after viability'[36]. In other words, it looks as if the Court's opinion is that there is

---

[35] Roe v Wade: https://caselaw.findlaw.com/us-supreme-court/410/113.html.

[36] 'With respect to the State's important and legitimate interest in potential life, the "compelling" point is at viability. This is so because the fetus then presumably has the capability of meaningful life outside the mother's womb. State regulation protective of fetal life after viability thus has both logical and biological justifications': p. 163 of ROE v. WADE Syllabus ROE ET AL. v. WADE, DISTRICT ATTORNEY OF DALLAS COUNTY APPEAL FROM THE UNITED STATES DISTRICT COURT FOR THE NORTHERN DISTRICT OF TEXAS No. 70-18. Argued December 13, 1971-Reargued October 11, 1972-Decided January 22, 1973. Pp. 113-178: https://tile.loc.gov/storage-services/service/ll/usrep/usrep410/usrep410113/usrep410113.pdf.

'potential life' up until there is 'fetal life after viability'. But, one may ask, what is the woman pregnant with? If she is pregnant with the 'potentiality of human life' then what is that potentiality except for viability? - for viability refers to the child being able to live outside the womb, with help, just as the womb is the right environment for the child to develop in. Thus the "Opinion of the Court" has conflated the existence of a child's life and viability whereas viability is only possible because of the existence of a child's life in the first place; and, therefore, the very recognition of viability is recognition of the life of a child from conception. The view that a child's life is not a child's life because it is not developed enough to exist outside of the womb is the same as claiming that any child that cannot live independently is not a child; but, in reality, it is in the very nature of being a child that development proceeds from conception onwards, through all the development stages that characterize a person's life, and which include pre-and-post birth stages of development. Thus the claim that a child is not a child if it is not possible for it to survive outside the womb is an incoherent and contradictory claim which has nothing to do with the reality of human development and everything to do with a prior judgement of a philosophical nature: a claim that is neither vindicated by biology nor logic. The claim that a child is not, by definition of his or her very existence, a human child, is contradicted by the very identity of the sperm and the egg through which he or she came to exist or, in the case of technological interventions, the human egg and nucleus through which he or she came to exist[37].

Furthermore, the claim that a child is not, by definition of his or her very existence, a human child, is not a logical claim; for, if what begins is typically human then he or she has a characteristically, developmentally, progressive expression of his or her identity. Therefore it follows that what has begun is typically the first of many stages through which his or her development passes in the course of showing "who" was present from the beginning; and, if this is true, then it follows that we are obligated to recognize the rights of this child: to life; to completing human development; and to integrity – to begin with but a few that need to be recognized. For, more generally, if the truth establishes our common human identity then it

---

[37] Cf. Profs. Justo Aznar and Julio Tudela: Chapter 5: Part II of *Conception: An Icon of the Beginning*, for a biological account of the human identity of the human embryo.

establishes, at the same time, our ethical responsibility for each other.

Potential life could be a quasi-philosophical concept which makes an unfounded distinction between being a potential human being and viability: as if viability introduces something that was not there before viability.

Alternatively, it looks as if the Court's opinion is that there is 'potential life' up until there is 'fetal life after viability'. Is, then, the "Opinion of the Court" saying that prior to 'viability' there is the potential for 'viability'? However, the potential for viability entails that a child is developmentally on course to becoming viable. It may be the case that 'viability' is, in the context of the "Opinion of the Court", relevant because the terms of the 14th Amendment refers to legal citizenship commencing, as it were, with birth (or naturalization); and, therefore, there may be the implicit claim that the 'viability' of the child brings him or her within the perceived jurisdiction of the 14th Amendment. For if the viability of the child is equivalent to surviving premature birth, being born prematurely nevertheless brings a person into American citizenship. Hence the implication that abortion is not the deliberate killing of an American citizen if the child is not 'viable' –

but only expresses a potentiality for viability outside the womb.

Viability, however, is a stage in the normal development of a child; and, therefore, the abortion of a child prior to his or her viability is clearly discriminatory and needs redress. According to the rule of law that redress could be according to how the right to life falls under the jurisdiction of a Constitution[38] or it could be more widely addressed on the basis of biolegal principles that can inform, not just national governments but the good of the human race[39]; indeed, in a certain sense, progress in the natural justice of a Constitution, or a law is progress towards the possibility of good international bioethical law.

In the end, then, aside from the possibility that viability is here being understood as determining whether or not

---

[38] From Ed Whelan, "Judge Barrett on Stare Decisis": https://www.nationalreview.com/bench-memos/judge-barrett-on-stare-decisis/.

[39] Cf. "Principles of international biolaw: Seeking common ground at the intersection of bioethics and human rights": Roberto Adorno, Bruylant, 2013: "Chapter 1: Principles of International Biomedical law", pp. 13-35: https://www.academia.edu/4063596/Principles_of_international_biolaw.

the Constitution can be said to apply to a viable child, there is also the possibility that there seems to be a coincidence of meaning between one reading of the 'potentiality of human life and the expression used in a document of the Catholic Church: that a woman is pregnant with a 'living being' that 'will be' a person or the woman is pregnant with whatever it is that has the 'potentiality of human life' but, again, is not yet a person.

In the first case there 'will be' a person and in the second case there is the 'potentiality of human life' and, presumably, at some indefinable point a human person. In the context of each document, however, there is a world of difference between how these two starting points are evaluated. In the case of *The Gospel of Life*, Pope St. John Paul II says: 'what is at stake is so important that, from the standpoint of moral obligation, the mere probability that a human person is involved would suffice to justify an absolutely clear prohibition of any intervention aimed at killing a human embryo' (*Evangelium Vitae*, 60). Whereas in the case of Roe v Wade the uncertainty as to what the woman is pregnant with has led to the possibility of the destruction of whatever she is pregnant with which has the 'potentiality of human life'. In other words, there is uncertainty in both expressions of what has come to exist at conception and yet they are evaluated in completely different ways.

What follows next is to discuss some aspects of the "Opinion of the Court" of Roe v Wade and the timely necessity of the development of Church teaching. Thus it is now necessary to examine, more closely, the issues that arise from an examination of the "Opinion of the Court" and a dissenting opinion in Roe v Wade.

CHAPTER SIX

ON THE INTERPRETATION OF TEXTS: PARTICULARLY THE 14TH AMENDMENT

AMENDMENT 14

'[1.] All persons born or naturalized in the United States, and subject to the jurisdiction thereof, are citizens of the United States and of the State wherein they reside. No State shall make or enforce any law which shall abridge the privileges or immunities of citizens of the United States; nor shall any State deprive any person of life, liberty, or property, without due process of law; nor deny to any person within its jurisdiction the equal protection of the laws'[40].

As, however, the 14th Amendment states that it is concerned with 'All persons born or naturalized in the United States, and subject to the jurisdiction thereof, are citizens of the United States and of the State wherein they reside' it could, therefore, be said to exclude a more universalist

---

[40] https://www.aclu.org/united-states-constitution-11th-and-following-amendments#14.

claim to apply to all human beings, whether in America by immigration or for whatever other reason. But then it also says: 'nor shall any State deprive any person of life, liberty, or property, without due process of law; nor deny to any person within its jurisdiction the equal protection of the laws'[41]. In other words, there is scope for the possibility that the meaning of the statement that no State can 'deprive any person of life' is in fact of a more universal character; and, therefore, there is a sense in which, perhaps not fully consciously, the legislators intended an implicit declaration of human rights which, in the nature of the law, is simply expressed according to the natural jurisdiction of this law applying specifically to those born or naturalized in America[42]. In other words, as citizenship

---

[41]   https://www.aclu.org/united-states-constitution-11th-and-following-amendments#14.

[42]   My perception of this point was sharpened by my daughter, Grace Etheredge's, reading of a draft of this article; and, in addition, the recognition that there will be a legal tradition defining the term 'person'. But, at the same time, it must be understood, too, that meaning is never confined to one expression of it and, therefore, there are wider considerations that

naturally applies to those 'born or naturalized in America', it follows that the American Constitution is not excluding the right to life from conception onwards; rather, the Constitution is defining the common understanding of an American citizen being, ordinarily, one 'born or naturalized in America'. Therefore, as we shall see in the following paragraphs, the right to life is within the general understanding of being 'born or naturalized in America' as, without life, there is no child to be 'born' an American citizen or, subsequently, no person to be naturalized.

What, then, will help with the interpretation of legal texts?[43] Thus, just as there is a concern for the original

---

are needed to help us to understand the use of any specific, legal terminology.

[43] See: "9 Things You Should Know About Supreme Court Nominee Amy Coney Barrett",

SEPTEMBER 28, 2020, by JOE CARTER:

https://www.thegospelcoalition.org/article/9-things-amy-coney-barrett/:

'5. In her judicial philosophy, Judge Barrett is considered a proponent of originalism, a manner of interpreting the Constitution that begins with the text and attempts to give that text the meaning it had when it was adopted, and textualism, a meaning of the philosophical understanding of 'first matter' and 'form'[44] which lies behind the triple conception of human being, whether in Aristotle or St. Thomas following him, so there is an interpretation of legal texts which endeavours to understand the original intention of the document. Just as there are principles and practices basic to the interpretation of Scripture which, generally, develop from the principle that the literal sense is the sense intended by the author[45], so there seems to be a parallel sense of interpreting legal texts: 'a method of statutory interpretation that relies on the plain text of a statute to determine its meaning'[46].

---

method of statutory interpretation that relies on the plain text of a statute to determine its meaning.

[44] 'Form' is here understood as that which determines the pure potentiality of 'first matter' to be a specific entity e.g. plant, animal or rational being.

[45] Cf. Etheredge, *Scripture: A Unique Word*, Cambridge Scholars Publishing, 2014.

[46] Joe Carter: 5th of "9 Things You Should Know About Supreme Court Nominee Amy Coney Barrett".

Clearly, then, the American Constitution follows Lincoln's address that 'all men are created equal'[47]; for, in the common understanding of being created equal, there is an ordinary sense, no doubt common at the time, that whatever constituted the original moment of a human being's creation was a moment common to all. Therefore, it could be said, whatever is now understood to be the common starting point of all human beings is what establishes our common equality before the law. Thus the principle: Where the body lives, there the soul is and where both are is the person[48]. Therefore, the first instant of human conception is ordinarily the first moment of a person beginning to exist. Where this first instant is subsequent to the first instant of fertilization, as in the case of twins, the first instant that the body of the twin comes to exist is the first instant that the twin human being has come to exist; and, irrespective of the difficulties of conjoined twins, the reality of conjoined twins shows forth the truth that bodily existence expresses personal existence. In the first instant

---

[47] Dr. Elizabeth Rex, "End Word", p. 607 of *Conception: An Icon of the Beginning*.

[48] See this theme developed throughout *Conception: An Icon of the Beginning*.

of what is established by artificial methods of human conception, it is the first instant in which the bodily existence comes to exist that there is what expresses the existence of the human person, whole and entire; it being the nature of bodily development to progressively manifest the existence of the human person from conception[49].

In other words, again in the common understanding of the time, a person would have been understood to exist from conception in that the life of the person exists from conception: parents conceive children not plants and mourn the loss of a child – not the loss of a plant; and, now

---

[49] Tina Beattie, "Catholicism, Choice and Consciousness: A Feminist Theological Perspective on Abortion", on p. 53: She says: 'I avoid scientific debates about embryonic development …' but then goes on to claim, on p. 60: 'There is no 'very first instant' of human existence, because the process between fertilization and implantation of the fertilized zygote takes several days.' However, without a 'first instant' there is no subsequent development; and, moreover, all subsequent development is continuous from the first instant. Therefore Beattie's claim is not only unscientific but incoherent. Personhood cannot be absolutely ruled out and, for the good reasons of this booklet, is thrice argued to be from the first instant of human conception: scientifically; philosophically; and theologically.

at conception is better understood to have a first instant,

en it follows that personhood begins from the first in-

nt of human conception: from the first instant that the

rm-egg union effects the whole of the embryonic child

ressed, bodily, in the enclosing of the sperm in the

ly formed and active embryonic wall. In other words,

e is a moment when there is an active sperm and an

egg and then there is the embryonic child expressed

e bodily integrity of the newly "walled" human em-

. For, quite apart from the philosophical problems

ave arisen and, in a sense, which have always accom-

d the definition of terms, there is the reality of a liv-

resence from the first instant of conception: a per-

life: what is understood to be the life of a human

: the life of a boy or a girl. The American Constitu-

then, establishes a wonderful precedent to which not

American abuses of this truth can be appealed[50], but

---

[50] Whether it be the abolition of slavery (Rex, "End Word" p. 607 of *Conception: An Icon of the Beginning*), the mistreatment of people suffering from syphilis (cf. Etheredge, *The Human Person: A Bioethical Word*, "General Foreword" by Dr. Mary Anne Urlakis, pp. 13-16) or the manipulation of ignorance which led to such tragic experiments on women in the

to which the world can "measure" its juridical claim to expressing and embodying equality before the law; indeed, while other legislation is more explicit, such as the 1990 German Embryo Protection Act passed 'in compliance with the Nuremburg Code'[51], the American Constitutional reference to the life of a person is as true now as it always was and communicates a common understanding that each one of us begins at conception.

Nevertheless, the opening wording of the 14th Amendment provides a context in which to understand the protection of the life of a person; for, it says: 'All persons born or naturalized in the United States, and subject to the jurisdiction thereof, are citizens of the United States'. In other words, a person is clearly understood to refer to 'All persons born or naturalized'; and, in general, a number of the cases which were heard with respect to this Amendment were concerned with, quite rightly, the reality of equality between races before the law[52]. Thus, it could be

---

trials of contraceptive pills (Etheredge, footnotes 780-781 on pp. 583-584 of *Conception: An Icon of the Beginning.*

[51] Pp. 597-598 of *Conception: An Icon of the Beginning*, Elizabeth Rex's "End Word".

[52] Cf. the "List of 14th Amendment Cases":

argued, the 14th Amendment is referring to the life of a person who is born, in that this simply reflects the concern of the legislation at the time, bearing in mind its desire to establish racial equality before the law; however, the life of a person who is born may well be a way of presuming that the life of an unborn child is implicitly defended in that this naturally leads to the life of 'All persons born'. In other words, the fact that the words of the 14th Amendment are concerned with the life of a person who is born is naturally the basis on which to recognize the implicit concern of protecting the life of all innocent people – from conception onwards.

One of the ways, then, of corroborating this claim is looking at the cultural climate which developed in the course of the period both leading up to and following the 14th Amendment. In other words, there are cultural, even legal developments, which elucidate the basic meaning of a text. In the case of the contemporary question of the

---

https://en.wikipedia.org/wiki/List_of_14th_amend-ment_cases; but, as one can see, there are many more cases that need to be considered as falling under this Amendment: https://en.wikipedia.org/wiki/Fourteenth_Amend-ment_to_the_United_States_Constitution.

validity of Anglican Orders, the question of what intention was expressed in the Anglican ordination of a priest,
it was historically demonstrated that an Anglican priest
was ordained in a specifically different sense to that of his
contemporary, Catholic priest. In other words, an Anglican ordination was specifically intended to repudiate, or
reject, the Catholic understanding of the priest as cele-
brating, effectively, a sacramental mystery of the presence
of the body and blood of Jesus Christ[53]. The point of this
observation, however, is not to confuse the issue by as-
suming that there is no other intention possible, either
then or now, but rather to recognize that a historical account can elucidate what is specific in a public enactment
characteristic of a historical period. Similarly, then, the

---

[53] Cf. For example, a concise summary of the point in
question raised and explained by Dr. Francis Clark, SJ, in *Anglican Orders and Defect of Intention*. By Francis Clark, S.J. Pp.
xx + 215. London: Longmans, Green, 1956. 25s.
   A. R. Vidler [a1]
   https://www.cambridge.org/core/journals/journal-of-ec-
clesiastical-history/article/anglican-orders-and-defect-of-in-
tention-by-clarkfrancis-sj-pp-xx-215-london-longmans-
green-1956-25s/F80F57967A7D266CC9F7F2EDBE849D92.

very developing of laws, across the United States[54], protecting the unborn child from the possibility of *abortion is itself confirming evidence of the historical intentionality expressed in the 14th Amendment*. Indeed, this is precisely the argument of the dissenting judge in Roe v Wade:

'MR. Justice Rehnquist, dissenting

'As early as 1821, the first state law dealing directly with abortion was enacted by the Connecticut Legislature' ... 'By the time of the adoption of the Fourteenth Amendment in 1868, there were at least 36 laws enacted by state or territorial legislatures limiting

---

[54] Cf. Christian Myers, "Law Professor Reflects on Landmark Case": 'Barrett also outlined the history of the Roe v. Wade decision and associated cases in the Supreme Court.

At the time of the case, most states prohibited abortion, except in cases wherein it protected the life of the mother, she said': https://ndsmcobserver.com/2013/01/law-professor-reflects-on-landmark-case/; and, more generally, the article by Steven Mosher: "How Amy Coney Barrett will use science and legal principles to overturn Roe v. Wade": https://www.lifesitenews.com/blogs/how-amy-coney-barrett-will-use-science-and-legal-principles-to-overturn-roe-v-wade.

abortion.'[55] In due course Rehnquist says: 'The only conclusion possible from this history is that the drafters did not intend to have the Fourteenth Amendment withdraw from the States the power to legislate with respect to this matter'[56].

In other words, both before and after the 14th Amendment the very development of legislation which protected the life of the unborn child, as it became increasingly threatened due to the increasing practice of abortion *is itself evidence of how the 14th Amendment's protection of the*

---

[55] Pp. 174-175 of pp. 113-178: ROE v. WADE Syllabus ROE ET AL. v. WADE, DISTRICT ATTORNEY OF DALLAS COUNTY APPEAL FROM THE UNITED STATES DISTRICT COURT FOR THE NORTHERN DISTRICT OF TEXAS No. 70-18. Argued December 13, 1971-Reargued October 11, 1972-Decided January 22, 1973: https://tile.loc.gov/storage-services/service/ll/usrep/usrep410/usrep410113/usrep410113.pdf.: the precise law was included in the text but here is in this footnote: 'Conn. Stat., Tit. 22, §§ 14, 16.'

[56] P. 177 of pp. 113-178 of ROE v. WADE Syllabus ROE ET AL. v. WADE etc.

life of the person was and is to be understood. In other words, while the 14th Amendment was specifically addressing the rights of citizens, the more universal right to life is implicated; and, as such, the 14th Amendment understood its remit to be that of referring to legal citizenship and, as such, no doubt had in mind the common practice of registering a birth, or the naturalization of a citizen, – as the natural point from which citizenship dates. Citizenship, however, registers a workable legal definition of a subject of the American Constitution; and, therefore, citizenship is the natural expression of what follows on a human life *as it becomes possible to identify the application of citizenship.* Thus, for example, a premature baby falls quite squarely within the meaning of the birth of an American citizen; and, even recently, it is furthermore recognized in President Trump's specific intention to sign a "The Born-Alive Infant Abortion Survivors Act" which will ensure medical treatment of a child that survives an abortion[57]. The question, then, of citizenship is

---

[57] Christine Rousselle, "Trump announces 'Born Alive' executive order for abortion survivors": https://www.catholic-newsagency.com/news/trump-announces-born-alive-executive-order-30262.

different from the question of preserving the life of a child; but, clearly, they are related: if the life of a child is not protected from conception then there is not a citizen to be legally recognized on being born and registered. The recognition of potential citizenship is different, then, to the speculative claim of a 'potential life': the former is a coherent legal definition whereas the latter is both unverifiable and derived, it seems, from a quasi-philosophical understanding of the relationship of a child's viability to a right to life.

On the Question of the Rightful Protection of Women

As regards the related claim that pro-abortion legislation is about

protecting women[58] it is true that, in general, regulating a procedure entails protecting a person from harm; and, indeed, in view of an abortion that is almost equivalent to self-harm, there is indeed the possibility of a great injury that the woman, or someone else, inflicts on her – but it is

---

[58] There are various possible references for this discussion, around pages 140-160 of 113-178 of ROE v. WADE Syllabus ROE ET AL. v. WADE etc.

not on her alone. Nevertheless, this tragic response to being pregnant is not answered by "medicalizing" abortion itself – as if it solely about protecting the woman. In other words, it is far from established that there is no harm to the woman, *per se*, in virtue of the very act of carrying out even a medicalized, or elective abortion. Indeed, the many testimonies of women who have had abortions and profoundly regretted them[59], quite apart from increasing psychiatric evidence of an adverse psychological reaction to what was thought to be some kind of simple "medical procedure"[60], not to mention the various medical complications of such an act[61], all contribute to the view that an elective abortion is not in the interest of the health of

---

[59] Why is this kind of evidence not taken into account? See: http://abortionmemorial.com/; indeed, it is increasingly recognized that the father, as well as other members of the family, are suffering too (cf. also: https://foundationsoflife.org/shes-pregnant-what-do-i-do-now/.

[60] Cf. "Post Abortion Syndrome": https://thelifeinstitute.net/learning-centre/abortion-effects/post-abortion-syndrome.

[61] Cf. "Risks about Abortion": https://foundationsoflife.org/facts-about-abortion/risks-about-abortion/.

the woman. Therefore, whatever distressing circumstances there may be involved in the actual context of a
woman deciding on an abortion, it is clear that an abortion is an invasive medical action on "who" is not part of
the woman's body; for, if the child were a "part of the
woman's body"[62] then the same relationship would exist
between her and her own body, whether it be a natural
part of her like a limb or a growth, like a cancer. But, as we
know, the loss of a limb is a different kind of grief to the
loss of a child. If, in other words, there is no child present,
then the medical action would be an amputation or the
removal of a random growth, and not an abortion. In a
word, then, the direct object of an elective abortion is the
removal of a child that a woman is carrying; and, at the

---

[62] 'In fact, it is not clear to us that the claim asserted by
some amici that one has an unlimited right to do with one's
body as one pleases bears a close relationship to the right of privacy previously articulated in the Court's decisions. The Court
has refused to recognize an unlimited right of this kind in the
past': p. 154 of 113-178 of ROE v. WADE Syllabus ROE ET AL.
v. WADE etc. Where is the proof that carrying a child falls under the terms of the following?: a 'right to do with one's body
as one pleases'.

same time, the removal of this child is generally done in such a way as that it does not involve the least consideration for preserving the life of that child. Deliberate abortion, then, for whatever reason, entails the implication of a "relationship" to the child, not just lost but removed, and this unaddressed relationship, whether denied or showing itself in unexpected ways, needs ultimately to be addressed by both the child's mother and father.

By implication, then, a medical act that intends to help the unborn child to be better placed for his or her development and, at the same time, is an act by which the mother is helped, is clearly of a different kind; and, therefore, one wonders why, given the increase in medical expertise, there are so few documented cases of a woman being helped with an ectopic pregnancy: a human embryo that implants in the fallopian tube or in some other place rather than in the womb. In other words, does the overriding mentality of abortion and abortifacient drugs obscure the real help that both mother and child need?

In sum, having read through the "Opinion of the Court" in Roe v Wade there is neither any real recognition of the harm to the woman of the event of the abortion itself nor any mention of the obvious harm to the child: the truth about what actually happens to a child in an elective abortion. However, reviewing the "Opinion of the Court" has, very helpfully, brought out a number of aspects that, as we can see, need a resolution.

Principle of Determining Appropriate Level of Legal Action

> 'Barrett ... states that she "tend[s] to agree with those who say that a justice's duty is to the Constitution and that it is thus more legitimate for her to enforce her best understanding of the Constitution rather than a precedent she thinks clearly in conflict with it. That itself serves an important rule-of-law value"'[63].

In a word, the historical consistency with which a statute is interpreted, whether in a constitution or elsewhere, reflects its underlying meaning and is a part of the evidence of what constitutes its original and stable meaning; and, therefore, the evidence of legislation, pre and post 14th Amendment, protecting human life against a rising tide of claims to justify deliberate abortion is itself

---

[63] From Ed Whelan, "Judge Barrett on Stare Decisis": https://www.nationalreview.com/bench-memos/judge-barrett-on-stare-decisis/.

evidence of what was understood by the 14th Amendment's protection of the life of the person. Furthermore, then, in the specific context of the American Constitution it makes sense that the first duty of a judge is to express 'her best understanding of the Constitution'; and, at the same time, it makes sense to have a view as to which level of the judiciary or the legislature is appropriate for the handling of a particular case or enunciating a particular law for the good of both *the rule of law* and *the good government of a country*[64]. More widely, then, this raises the possibility of biolegal principles that can inform, not just national governments but the good of the human race[65].

---

[64] MICAIAH BILGER: Amy Barrett Believes Life Begins at Conception, Questions Roe's "Judicial Fiat" of "Abortion on Demand":

https://www.lifenews.com/2020/09/25/amy-barrett-believes-life-begins-at-conception-questions-roes-judicial-fiat-of-abortion-on-demand/.

[65] Cf. "Principles of international biolaw: Seeking common ground at the intersection of bioethics and human rights": Roberto Adorno, Bruylant, 2013: "Chapter 1: Principles of International Biomedical law", pp. 13-35: https://www.academia.edu/4063596/Principles_of_international_biolaw.

CHAPTER SEVEN

AN ANSWER TO THE UNCERTAINTY OF WHAT OR WHO EXISTS FROM CONCEPTION

Uncertainty is a characteristic of human experience; and, therefore, it is not unusual for a variety of factors to help determine the reality of what actually exists: What the real situation actually is. Uncertainty, however, expresses a value. There is an uncertainty about whether or not a specific act of spousal love will beget a child; and, indeed, it is possible that the very uncertainty that exists allows for the perception and reception of a child as a gift: a gift from God[66]. In the case of pregnancy, then, these two very different documents agree that there is an uncertainty about what is happening. On the one hand the woman is pregnant with a 'living being' and, on the other hand, with what has a 'potentiality of human life'. In neither case does there seem to be any clarity about what comes to exist at conception; except, that is, there is a common agreement that "something" comes to exist at

---

[66] Cf. Etheredge, *The Human Person: A Bioethical Word*, St. Louis: En Route Books and Media, 2017: pp. 61-62.

conception. The question is, then, what comes to exist at conception. What follows are a number of considerations which, together, constitute an answer to the need for certainty.

What is the Experience of Women in Pregnancy?

My wife spoke of looking forward to meeting the person conceived[67]; and, as such, echoed the certainty of Eve: 'I have gotten a man with the help of the Lord' (Gn 4: 1; but consider the whole biblical witness to the real experience of women, children and men written about in the Scriptures). Even, then, if it was not so clear to my wife that conception involved an act of God, it was certainly clear to my wife that she had conceived a child: 'a man' (Gn 4: 1). There are many other testimonies, too, both in

---

[67] This was put more formally in the article, Francis Etheredge, "The Mysterious Instant of Conception": *The National Catholic Bioethical Quarterly*, Autumn 2012:

https://www.pdcnet.org/C1257D43006C9AB1/file/11DA
AEF87F30A61985257D6D00682134/$FILE/ncbq_2012_0012_
0003_0041_0050.pdf.

human experience generally[68] and in the human experience embedded in the Scriptures. What is the identity of the human longing for a child which makes the suffering of infertility so painful?[69] What is so disappointing about a miscarriage? Surely it has nothing to do with the abstract claim of losing a blob of cells: a claim so contrary to the reality of the organized human embryonic child whose development is so interactively ordered to his or her presence in the nurturing womb of his or her mother[70]. What

---

[68] Cf. Etheredge, *The Prayerful Kiss,* St. Louis: En Route Books and Media, 2019: the poem and prose entitled "Indelible": an account of the loss of a child to abortion from the experience of a father.

[69] Cf. Etheredge, *Mary and Bioethics: An Exploration,* St. Louis: En Route Books and Media, 2020, See Leah Palmer's "Foreword to Chapter Seven" pp. 223-228 and Chapter Seven: "Love, Scripture, Suffering and Bioethical Questions", pp. 229-254; and see Adriana Vasquez's Foreword, pp. 84-91, to Chapter Two: "Marriage is a Liturgical Act" of Etheredge's *The Human Person: A Bioethical Word,* 2017.

[70] Cf. Etheredge, *Conception: An Icon of the Beginning*: the following is a profound analysis of the evidence by Profs. Justo Aznar Lucea and Julio Tudela: Chapter 5: Part II: "The

is the value of this human experience? Why is it only a question of arguments based on often difficult philosophical positions when, in reality, there is a wealth of human experience, particularly the experience of the woman and mother?[71]

It is true, however, that there are many and varied injustices in the treatment of women and, in particular, in the abuse of women that leads to pregnancy; however, these problems need addressing independently of the child that may be conceived as a result of this mistreatment, just as children need help in families where the parents are struggling in other ways.

The Witness of Each One of Us

In contrast to the uncertainty that surrounds human conception, there is no doubt that a person who comes to

---

Biological Status of the Early Human Embryo, When Does the Human Being Begin?", pp. 480-507.

[71] Cf. Etheredge, *Conception: An Icon of the Beginning*, for Scriptural citations under the heading: Chapter Two: "The woman's perception of conception: Part V of XII", pp. 158-162 and drawing further on Job up to p. 166.

exist, comes to exist at a certain point in time. In other words, each one of us is an indelible, irrevocable and incontrovertible witness to the three-dimensional fact: firstly, each one of us comes to exist; secondly, we come to exist amidst multiple relationships, beginnings with those from whom we received our ordinary human inheritance, ordinarily our parents; and, thirdly, that our life is a gift and, if we are well disposed to recognizing it, we recognize it as a gift. Whatever we may think about the multitude of questions that surround human conception, that we are conceived is certain: that conception, which means 'beginning', is certain. The contrary claim, that we did not come into existence, is clearly contrary to the facts of the union of sperm and egg and the existence of each one of us. Even with respect to the manipulation of human life which involves the multiple injustices of fertilization in a glass dish, discarding unwanted human embryos, freezing of human embryos, experimenting on human embryos and combining different paths to the whole of conception, wherever there is the living human body there is the presence of the person. Otherwise why would husband and wife, from the dawn of time, come together and exclaim, like Eve, 'I have gotten a man with the help of the Lord' (Gn 4: 1)? Otherwise why would a human egg be fertilized in a 'petri dish' in 1977, and born in 1978, make the birth of Louise Brown 'headline news around the world'[72]? Otherwise why would a frozen human embryo returned to the nurturing womb be a human being: a girl called Hannah?[73] Thus there is the corresponding human right of the human life, once conceived, in whatever way he or she is conceived, to the completing nurture of maternal implantation and development.

A Discussion on the Teachings of the Catholic Church and the "Opinion of the Court"

In the course of the "Opinion of the Court", although there was a review of ancient views on human conception, the judge said that it is now the 'official belief of the Catholic Church' to 'recognize the existence of life from the

---

[72] Dr. Elizabeth Rex, "End Word" on p. 596 of Etheredge, *Conception: An Icon of the Beginning.*

[73] Dr. Elizabeth Rex, "End Word": '1998 – Hannah Strege, the world's first adopted frozen embryo is born in San Diego, California on December 31, 1998' on p. 599 of Etheredge, *Conception: An Icon of the Beginning.*

moment of conception'[74]. But, having said that, it appears that St. Thomas Aquinas' antiquated biology, among other sources, was used to justify the claim that ordinarily there is not a human being from the first instant of conception[75] because, in the "Opinion of the Court", the judge said: 'We need not resolve the difficult question of when life begins. When those trained in the respective disciplines of medicine, philosophy, and theology are unable to arrive at any consensus, the judiciary, at this point

---

[74] Cf. Pp. 160-161 of pp. 113-178 of ROE v. WADE Syllabus ROE ET AL. v. WADE etc.

[75] Dr. Elizabeth Rex: "End Word": '1973 – On January 22, the U.S. Supreme Court legalizes abortion in Roe v. Wade. The majority decision uses 13th century theology and science to support their erroneous decision about when human life begins, stating: *"Christian theology and canon law came to fix the point of animation at 40 days for a male and 80 days for a female,* a view that persisted until the 19th century.... *Due to continued uncertainty about the precise time when animation occurred, to the lack of any empirical basis for the 40-80 day view, and perhaps to Aquinas' definition of movement"* (footnote 807: Roe v. Wade, 410 U.S. 113 (1973) IV.3, with emphasis and italics added) on p. 611 of Etheredge, *Conception: An Icon of the Beginning.*

in the development of man's knowledge, is not in a position to speculate as to the answer'.[76] Firstly, then, the
statement of the judge clearly neglects to consider his own
observation that Catholic Teaching, resourced as it is by
numerous experts, now teaches 'the existence of life from
the moment of conception'. Secondly, the judge showed
signs of recognizing the growth of a consensus when he
said: 'As one brief amicus discloses, this is a view strongly
held by many non-Catholics as well, and by many physi-
cians'[77]. Thirdly, the embryological evidence which
would have helped the judge recognize the beginning of
human life resulted in the opposite, namely, misleading
him to think that there was no clarity about the beginning
of human life; and, therefore, as an objection to the view
of a first instant of human conception the judge said: 'new
embryological data ... purport to indicate that conception
is a "process" over time, rather than an event'[78]. In other

---

[76] P. 159 of 113-178 of ROE v. WADE Syllabus ROE ET
AL. v. WADE etc.

[77] P. 161 of 113-178 of ROE v. WADE Syllabus ROE ET
AL. v. WADE etc.

[78] P. 161 of 113-178 of ROE v. WADE Syllabus ROE ET
AL. v. WADE etc.

words, there is no inconsistency between conception as a process over time and a first instant of fertilization. For, a process over time has a beginning, namely that there is a first instant that the sperm is enclosed by the closure of what was the egg's open pores and thus the walled embryo is the first stage of a new entity: the nascent human being; this first instant of the human embryo is in contrast to the prior, separate existence, of the active human sperm and the inert human egg. Furthermore, then, the process of fertilization is simply the first stage of human growth from the first instant of fertilization to the fusion of the respective nuclei of what was a sperm and an egg and the expulsion of unrequired chromosomes. Normal development, as even the Warnock Report concluded, proceeds uninterruptedly from the beginning onwards[79]. Fourthly, the

---

[79] It is necessary to note that even the Warnock Report recognized that there is a seamless process of development: 'there is no particular part of the developmental process that is more important than another; all are part of a continuous process' (Department of Health and Social Security [UK], *Report of the Committee of Inquiry into Human Fertilisation and Embryology* (London: Her Majesty's Stationery Office, July 1984), para. 11.19, quoted in Catholic Bishops' Joint Committee on Bio-

presence of 'new medical techniques such as menstrual extraction, the "morning-after" pill, implantation of embryos, artificial insemination, and even artificial wombs' seemed to have confused the issue of when life begins because each of these medical techniques required a careful examination of their relationship to the beginning of life and, therefore, should have at least raised a cautionary note about their use as an objection to there being a moment of the beginning of human life. Indeed, as already noted, where the human body lives there is the presence of human personal life – especially bearing in mind the principle that biological development is inherently, as it were, psychological development: being a male or female

ethical Issues, *Response to the Warnock Report on Human Fertilization and Embryology* (London: Catholic Media Offices, 1984), 13. In response, the Catholic Bishops of Great Britain said that 'our society should resolve to protect the life of the human embryo *precisely from the beginning of its continuous development, ie, from conception (fertilization)*' (Catholic Bishops' Committee, *Response to the Warnock Report*, 13). In other words, the very truth of the seamless process of development requires the welcome of each human life from conception.

person is to be a 'psychologically inscribed embryologically-begun human individual'[80].

Finally, then, the following two claims are problematic: that 'We need not resolve the difficult question of when life begins' and 'When those trained in the respective disciplines of medicine, philosophy, and theology are unable to arrive at any consensus, the judiciary, at this point in the development of man's knowledge, is not in a position to speculate as to the answer'[81]. They are problematic because, in effect, the "Opinion of the Court" has speculated about when life begins, calling it 'potential life', and has concluded that the state need not defend the interest of the nascent human life from his or her very beginning. In other words, instead of the court ruling that the question of the beginning of human life required further investigation and withdrawing from making a judicial judgement, the court implicitly held a quasi-philosophical view of

---

[80] Etheredge, *The Human Person: A Bioethical Word*, p. 311.

[81] P. 159 of 113-178 of ROE v. WADE Syllabus ROE ET AL. v. WADE etc.

there being a 'potential life' and permitted the destruction
of a child from his or her beginning.

A Clarification as Regards the Teaching of St. Thomas
Aquinas

As regards the view of St. Thomas Aquinas; he argued,
not simply that there are three stages to ordinary human
development but that, as a whole, nature intends a man:

> 'St. Thomas embraced such a comprehensive account
> of Christian and philosophical thought that it is worth
> considering his understanding of the implication of the
> delayed ensoulment of a human being; he said: 'foe-
> tuses are animal before they are human ... [but] nature,
> in producing the animal foetus, is aiming at producing
> a man'[82].

What is more, the philosophical biology on which St.
Thomas based his three phase conception of human being

---

[82] *Summa Theologiae*, Methuen, Pt I, Qu 85, art 4, p. 136;
but quoted from Etheredge, *Conception: An Icon of the Begin-*
ning, p. 237.

was Aristotelian in that Aristotle held that matter was eternal and that, therefore, form (which determines what first matter will be) was always necessary to differentiate the matter that eternally existed into its various kinds; and, as it was held that human conception was not developed enough to receive a human soul, so it was understood that there was a three phase ensoulment: plant; animal; and then rational ensoulment[83]. But, all the while, St. Thomas held the sophisticated view that nature, nevertheless, 'is aiming at producing a man'; and that, therefore, even modern embryology confirms the view that the fruit of spousal union, ordinarily, in fact and from conception intends a man or a woman.

What seems to be much less well known, is that St. Thomas Aquinas also argued that the conception of Christ was immediate: 'conception of the body [does not precede] ... animation by a human soul ... in Christ'[84];

---

[83] For a more comprehensive discussion of this go to Etheredge, *Scripture: A Unique Word*, Cambridge Scholars Publishing, 2014, pp. 303-306.

[84] *Summa Theologiae*, Methuen, Pt III, Qu 5, art 5, p. 484, footnote 239 of p. 233 of Etheredge, *Conception: An Icon of the Beginning.*

and, indeed, it is this understanding, explicit in St. Maximus the Confessor, which advances the view that what happened to Christ, notwithstanding the virginal conception, is what happens to us. 'Fr. John Saward, drawing on St. Maximus the Confessor, says: 'Apart from the saving novelty of its virginal manner, the conception of Christ is in all respects like ours'[85]. In other words, while there is an argument from St. Thomas concerning the delayed animation of a being from conception by a rational soul – there is a more important argument that Tradition has taken up, namely, that human conception follows, however imperfectly, the conception of Christ: that just as Christ was one in body and soul from conception so we are one in body and soul from conception.

The Contribution of Revelation and Dogma

Drawing 'on St. Maximus, Fr. Saward says that 'if the embryo immediately after fertilization is endowed with only a vegetative soul, then men father plants, not men.

---

[85] In footnote 249 of *Conception: An Icon of the Beginning: Redeemer in the Womb*, Ignatius Press: San Francisco, 1993, p. 12, but see also pp. 8-13.

But in fact the act of fertilization establishes a human-to-human relationship between father and child; *I am conceived by my father*"[86]. Alternatively, what is the living human life which has begun: Is it not an actual human life with the potential of manifesting the whole presence of the human person? Within the very tradition of the Church, then, there is the unique conception of Christ drawing us ever closer to the truth of human conception, one in body and soul from the first instant of human existence; and, in addition, there are scriptural and other dogmatic resources to be drawn upon to elucidate the mystery of human conception. Indeed, notwithstanding the different accounts of human conception and their authors' purposes[87], there is a sense that who is conceived is conceived as a whole; as David says:

---

[86] Etheredge: *Conception: An Icon of the Beginning*, p. 221, quoting from *Redeemer in the Womb*, Ignatius Press: San Francisco, 1993: p. 10, drawing on the Ambigua 2, 42; 1337B-1340B.

[87] In Genesis, for example, there is a very different account to the creation of human being, male and female, to that expressed elsewhere and yet its very difference is open to interpretive explorations, albeit not contradicting natural truths (according to St. Augustine); see "Chapter Two: Scripture and the

'Thy eyes beheld my unformed substance; in thy book were written, every one of them, the days that were formed for me, when as yet there was none of them' Psalm 139: 16; and, indeed, the very unique Hebrew word, *golmi,* communicates an incredible summary sense of 'an unfinished vessel' which is inescapably personal: my unfinished vessel[88].

At the same time there is the mystery of Mary, the Mother of the Lord which has, in the last century, come into great and greater prominence for a variety of reasons but, in truth, to assist us in a timely and providential understanding of both the mystery of the Church, the mystery of salvation in Christ and the mystery of each one of us. This great untapped mystery offers us a singular glance at the moment of human conception, primarily so that we can understand that Christ inherits human flesh free from original sin – but nevertheless it does so in such a way as to illuminate the moment characteristic of human conception:

Beginning of Human Being", *Conception: An Icon of the Beginning.*

[88] Cf. pp. 190-200 of *Conception: An Icon of the Beginning.*

'If grace requires the presence of the human soul, then for grace to be effective in the flesh, as it were, as well, then body and soul need to be united. Thus the mystery of the *Immaculate Conception* implies that Mary is one in body and soul (*Gaudium et Spes*, 14) at the instant of their reciprocally coming to exist; indeed, as it says simply in *Lumen Gentium*: 'Enriched from the first instant of her conception with the splendor of an entirely unique holiness, the virgin of Nazareth is hailed by the heralding angel, by divine command, as "full of grace" (cf. Lk. 1: 28 ...)' (56). In other words, while the Church does not explain the 'first instant of conception' – the ultimate 'first instant' is the first instant that the sperm animates the egg and the embryo expresses this through the formation of the embryonic wall'[89].

In a word, then, just as St. Thomas Aquinas argued that we need the help of Revelation to aid our understanding of whether or not there was a beginning to creation so we

---

[89] Etheredge, *Mary and Bioethics: An Exploration*, p. 166; and the footnote to this quotation goes to Etheredge, Chapter 12, *Scripture: A Unique Word*, in which the intricate evidence for these claims is discussed.

need the help of Revelation to determine the truth concerning human conception: a truth which confirms and expresses the common understanding of personal experience.

A Variety of Bioethical Declarations

Controversies, as we know, have raged and will continue to rage down the centuries of human history; but, in contrast, we have a number of helps. The Hippocratic Oath states: 'I will not give a woman a pessary to procure abortion'[90]. The Nuremburg Code says: 'No experiment should be conducted where there is an *a priori* reason to believe that death or disabling injury will occur'[91]. The Belmont Report says: 'persons with diminished autonomy are entitled to protection'[92]. Indeed there are any number of wonderful declarations that seek to draw the truth from

---

[90] Courtesy of Dr. Mary Anne Urlakis, p. 21 of *The Human Person: A Bioethics Word*.

[91] https://media.tghn.org/medialibrary/2011/04/BMJ_No_7070_Volume_313_The_Nuremberg_Code.pdf.

[92] https://www.hhs.gov/ohrp/regulations-and-policy/belmont-report/read-the-belmont-report/index.html.

human experience, rectify wrongs and establish a way forward for us all; and, indeed, as it has been discussed, it involves the use of technical terminology which, however, has an understandable significance: the reality that each one of us begins: a beginning which entails an inviolability which requires recognition and, where appropriate, the remedies of medical help which are possible and applicable for the benefit of each human subject, whether embryonic or adult. In a word, however, there is an ongoing necessity that there be an explicit recognition of what constitutes both the historical truth of what was intended by specific national legislation and its updating according to a more explicit understanding of a relevant reality, such as human conception; but, also, we are in a new human context that requires a renewed understanding that even specific, national laws, exist in the context of universal truths and rights and that, in the end, these are a part of what will fashion the future for all of us. Thus, it seems, we are already beginning to see the articulation of principles which are capable of being the basis of national and, possibly, international law: 'it is clear that the principle of respect for human dignity, due to its non-negotiable character and overarching scope, will always play a crucial role in every decision concerning biomedical practice'[93].

Gravitating to a Consensus

Without a beginning there cannot be development and without development there cannot be a precise perception of what began; but, on the basis of what unfolds from a beginning, so what began from that beginning is made manifest. In the words of Pope Francis, collaboratively expressed with the Grand Imam Ahmad Al-Tayyeb, "'In the name of innocent human life that God has forbidden to kill, affirming that whoever kills a person is like one who kills the whole of humanity, and that whoever saves a person is like one who saves the whole of humanity …'[94].

---

[93] "Principles of international biolaw: Seeking common ground at the intersection of bioethics and human rights": Roberto Andorno, Bruylant, 2013: "Chapter 1: Principles of International Biomedical law", p. 19 of pp. 13-35: https://www.academia.edu/4063596/Principles_of_international_biolaw.

[94] Pope Francis, *Fratelli Tutti*:

http://www.vatican.va/content/francesco/en/encyclicals/documents/papa-francesco_20201003_enciclica-fratelli-

How true this is turning out to be when whole countries permit the destruction of the unborn, not to mention international organisations which promote abortion, pharmaceutical and IVF companies which have no regard for the truth that human life begins at the first instant of conception and that each person has a right to an integrally human identity and his or her completing human de-

---

tutti.html#_ftnref262. In *Conception: An Icon of the Beginning*, there is Muslim scholar who holds the view of human life from conception, p. 575: 'Germany, however, has gone before the world and enacted the following, 1991 legislation: "Act for the Protection of Embryos" (The Embryo Protection Act). This "word" of law, as it were, has become a world-wide teacher and a noted Muslim bioethicist commented, approvingly, on the German law: Hassan Hathut (1924-2009) 'referred to Germany, which banned all use of human embryos in biomedical research. As for the surplus of fertilized ova in the IVF processes, the law even banned initiating such a surplus …. Hathut concluded that this law goes in line with Islamic ethics (Hathut 1994, 175)' (citation on the Muslim scholar from: "'Islam, Paternity, and the Beginning of Life": "The Beginning of Human Life: Islamic Bioethical Perspectives" with Mohammed Ghaly, (*Zygon: Journal of Religion and Science*, vol. 47, No. 1, (March 2012), pp. 175-213: https://core.ac.uk/download/pdf/43497555.pdf, p. 207.'

velopment. By contrast, however, there is a growing international recognition of these human rights which are beginning to be reflected in what is called the "Geneva Consensus Declaration": the Center for Family and Human Rights 'has worked for 24 years toward the declaration made by the Trump administration today together with a coalition of 32 UN Member States. There is no international right to abortion. There is no international obligation to fund abortion. The United Nations has no business interfering in sovereign decisions when it comes to protecting life in the womb' (October 22, 2020)[95].

This is not a naïve proposal[96] – but would clearly require patient, persistent and prudent international co-

---

[95] First paragraph of "Statement of Austin Ruse, President of C-Fam, on the signing of the Geneva Consensus Declaration" (October 22, 2020): https://email.opusfidelis.com/t/ViewEmail/j/92252BD7C1C8076B2540EF23F30FEDED/80B27ACA6DF63276F7E8006BBCB98688; cf. also: Austin Ruse, 23 October, 2020, "Governments Launch Pro-Life Declaration at United Nations: https://c-fam.org/friday_fax/governments-launch-pro-life-declaration-at-united-nations/.

[96] A relevant essay by Teresa Etheredge, and an article by Grace Etheredge, have helped me to think through these

operation. As Roberto Andorno has said: 'Global challenges raised by biomedical advances require global responses'[97].

---

questions: Teresa Etheredge: "Why is the criminological imagination important to the future of criminology? Describe a current issue in criminal justice and explain how the criminological imagination could help us to understand it"; and Grace Etheredge: "The impact of public international law on UK courts".

[97] "Biomedicine and international human rights law: in search of a global consensus": "Abstract": p. 960 of the "Bulletin of the World Health Organization 2002, 80 (12)".

CONCLUSION

LEST WE FORGET MOTHER, CHILD AND FATHER

There will no doubt be ongoing controversies about the natural right of members of the human race to be conceived free of animal-human hybrid experiments[98] and the right to integral, completing human development: so that once a child is conceived, he or she can be protected from exploitation and even rescued from the frustration of being frozen. and, finally, there is the whole field of experimentation on embryonic human beings and the vested interests of investigators, multi-nationals, har-

---

[98] In a brief correspondence with Roberto Andorno, he said: 'But I don't think that human rights can really help to address the kind of issues that you mention (for instance, a right not to be conceived as a genetically modified human being) because human beings who do not exist yet -who have not even been conceived yet- cannot have any human rights... We need some other legal-conceptual tools that relate, not to so much to existing individuals, but to the integrity of future generations, and the "human condition" in general'
(https://www.academia.edu/Messages?atid=18910888).
Thus this raises the question of not just the future of the human race – but the human race as the real and substantial "subject" of human rights.

vesting of organs and a whole, almost unimaginable world of exploiting the human being as a "resource" – for this human being is our brother or sister. We stand, then, at a point in human history where it is not so much a question of personal choice determining anything and everything as choosing the truth, as it becomes more fully known concerning human conception, that will take us into a humane future of the human race or the future of the human race will be determined by the most powerful and prevailing vested interests that will determine, on utilitarian grounds, whose future it will be to be a resource for the rest of the human race. If there are documents explaining the ethical relationship to one another, because of our equality as human beings[99] and the tragic events of our times, then how much more necessary is a revisiting of these foundational expressions in the light, or flickering light, of the times in which we currently live.

In the end, then, this is not an abstract discussion although, at times, it takes us into the most difficult philo-

---

[99] Cf. Chapter 5: Part II, a masterful review of the biological evidence concerning conception written by Profs. Justo Aznar Lucea and Julio Tudela in *Conception: An Icon of the Beginning.*

sophical terminology there is, the most amazing and detailed analyses of embryology and the most intense and controversial social disagreements in modern times; it is a discussion about specific people, whether born or unborn: it is about who has received the gift of human life is simply equal to anyone else who has received the gift of life. Thus no one is excluded from the world discussion of what we, as human beings, are bringing about in the present ethical climate of the human race. As we emerge, then, from our national identities and increasingly recognize that abstract truths about human personhood, that to be a human person is to be a human being-in-relation, need "returning" as it were to the concrete reality from which they came – we will appreciate more and more that parent and child, brother and sister, aunt and uncle communicate the profoundly interpersonal structure of human identity.

We have discussed various ways to understand human being, drawing on analogies or comparisons so that what is unfamiliar, becomes intelligible and familiar; but, in the end, as we make actual progress in understanding the true reality of human conception, so we can see that our comparisons have strengths and weaknesses and that we are called, more and more, to communicate the truth that we have discovered: the truth to be "embodied" in a universal declaration of human rights. Thus, there is a mother who needs help [100]. There is a child who needs help. There is a father who needs help to understand his fatherhood [101]. And there is a world of people with vested interests who need help to appreciate that all human beings are a gift, equally given into the care of each of us; and, therefore, that whatever the good intended, there is an obligation on everyone to understand that there is an objective good for each and every one of us, without exception, otherwise there is an actual inequality between us. And if there is an actual inequality between us, as regards who is a human being, then there is the increasing possibility of an exponential increase in the exploitation of all of us; for human rights, in the end, are the rights of relationship: the rights of relationship which come into existence when human beings are conceived – conceived in relationship to the whole human race and to God.

---

[100] Myers, "Law Professor Reflects on Landmark Case": https://ndsmcobserver.com/2013/01/law-professor-reflects-on-landmark-case/.

[101] Cf. Etheredge, *The Prayerful Kiss*, particularly: "Indelible".

Just, then, as a plant exists in an ecological context and
what is done to it, for better or worse as regards the health
of the individual plant and its place in the ethico-eco-sys-
tem, so what is done to individual human beings impacts,
for better or worse, on the "ethical" whole of the integrity
of the human community. But, 'Conscious' as we are of
our 'limitations', we ask 'for constant prayer and interces-
sion for' our 'efforts to translate the sacred texts "in the
same Spirit by whom they were written"'[102]; and, therefore, we need to ask for the same help to read both the
book of nature and all that helps us to recognize what is
universally good, true and applicable to all. Thus, in a
spirit of universal fraternity, any progress in the Church's
teaching on conception can only complement any development in the natural understanding of it; and, in that
same spirit, one source can complement another while being very different from it. Thus, in the end, there is the
progress of the hope that international agreements and

---

[102] Pope Francis, *Scripturae Sacrae Affectus*, commemo-
rating the life and work of St. Jerome, citing *Praefatio in Penta-
teuchum*: PL 28, 184': http://w2.vatican.va/content/fran-
cesco/en/apost_letters/documents/papa-francesco-lettera-
ap_20200930_scripturae-sacrae-affectus.html#_ftn19.

laws will not only help the plight of people in the present but will lead to the protection of our common humanity; indeed, is it time that the "human race" becomes the "subject" of human rights and international agreements and, ultimately, international bioethical law?[103]

---

[103] Elective abortion is by definition discriminatory against the life of the child aborted; and, therefore, if the law progresses through the rejection of what is discriminatory, then it is right to reject elective abortion; cf. also "Arguments from International Human Rights Law in the Supreme Court's Dobbs v. Jackson Women's Health Organization Abortion Case" by Alexis I. Fragosa, Esq: https://c-fam.org/definitions/arguments-from-international-human-rights-law-in-the-supreme-courts-dobbs-v-jackson-womens-health-organization-abortion-case/.

POSTSCRIPT

ROE V WADE: ONGOING ARGUMENTS
OF BENEFIT TO US ALL

As the discussion continues about the relationship between the American Constitution, those who object to
abortion and those who advocate abortion, there is
emerging a certain amount of clarity around a number of
central arguments which, indeed, have a world-wide rele-
vance to the debate concerning the legal recognition of the
right to life of the living.

A general point, however, before moving on is the cir-
cumstantial situation of women in society and, depending
on the immediate culture, there may be more or less support for even a welcome pregnancy, never mind one that
is ether unexpected or even forced upon the woman. In
other words, there are always wider issues than that of the
immediate question of the right to life; and, in justice,
none of them can be neglected to the detriment of the
whole provision of help to family life[104]. In particular,

---

[104] It is a gross oversimplification to identify the Catholic
Church's response to abortion and the whole range of human
rights violations with one or two documents of the Church

and admittedly in the specific context of America, there are many significant changes in society, all of which need their own analysis and synthesis, but one of them is the rise of the number of men and women in prison: between a steady population of 10,000 at the turn of the century it has risen to $1,414,162$, in 2018[105]. In other words, during the 60s there began a trend which has only increased, namely, that just as natural law has become obscured and

---

when there are so many religious orders, priestly works and lay responses that, in justice, need to be recognized; cf. Beattie, "Catholicism, Choice and Consciousness: A Feminist Theological Perspective on Abortion", p. 57.

[105] Unpublished text supplied by the author and with permission to cite: Soozi Scheller, (with Dr. David S. Crawford, Law, Family, and the Person), "Unfruitful Law: The Woman as a Revealer of the Heart of the Culture", 23rd April, 2021, p. 6: 'Between 1925 and early 1960s, the population in the U.S. state and federal prisons had remained at about 10,000 men and women. However, by 1964, this figure had increased to 20,000 and further by 40,000 in 1984. With each subsequent decade, the increase became more rapid, so that in 2018 the number of incarcerated individuals reached 1,414,162, which is the highest per capita rate in the world' (footnote in the original to "Trends in US Corrections" etc).

marginalised so has the social transgression of the law of the land increased[106]. One wonders, then, more widely, how socially disorientating is the loss of reason's guide to moral action[107].

There are the following seven sections to this postscript: Justice Beyond a Change of Justices (i); Viability is for Life (ii); Choice, Burdens and their Alleviation (iii); Bodily Integrity, Liberty, Equality and the Constitution (iv); Brain Death and Abortion (v); Abortion and the Advancement of Women (vi); True Justice is Irreversible (vii).

Section i: Justice Beyond a Change of Justices. If whether or not a court decides to grant a legal right to

---

[106] E.g. The rise of hormonal contraception (which obscures the key woman's health indicator of ovulation), abortion and confusion about male and female identities); my inference but on the basis of some of Soozi Scheller's essay, "Unfruitful Law" etc.

[107] Cf. The forthcoming work, *Human Nature: Moral Norm*, by Francis Etheredge and from En Route Books and Media, maybe late 2022.

abortion depends on who has appointed the judges[108], then there is a problem with the criteria by which a legal decision is made; and, as such, it turns the question as to who lives and dies into an absolute lottery – for all who are the legally unrepresented subjects of a decision to allow any kind of deliberate abortion, experimentation or otherwise infringement of the right to protected continuing human development. In other words, the very nature of what makes these variable decisions possible is already demonstrating to the nation and the world that politicians and 'the Court ... [are] playing politics'[109]. But the point, surely, is not whether or not the Court or others are 'playing politics' but that there is a public duty for a court to determine what is just and just for all; and, therefore, both the Court and politicians cannot renounce their responsibility, before God and their nation, to formulate just laws.

---

[108] Cf. Mark Sherman, November 29, 2021, AP News: "Supreme Court set to take up all-or-nothing abortion fight".

[109] Helen Alvaré, December 1st, 2021 : "Supreme Court's 'Dobbs v. Jackson' Oral Argument" etc: https://www.ncregister.com/commentaries/supreme-court-s-dobbs-v-jackson-oral-arguments-promising-for-pro-life-cause.

While there are those who think that an outcome of the current debate could be a 'limited "right to abortion"'[110], Justice Clarence Thomas says that there is not a 'shred of support' for a right to abortion in the American Constitution[111]. In view, then, of the clear statement of the 14th Amendment, that no State shall 'deprive any person of life, liberty, or property, without due process of law' – it is possible that a review of what the Constitution says will have to address the obvious contradiction between a supposed 'right to abortion' and the prohibition against depriving anyone of life. At the same time, 'Some of the briefs ask the Court to let each state decide abortion law for itself'[112] – but again this does not go beyond the

---

[110] David Bjornstrom, Esq, "Dobbs Case Exposes" etc: https://personhood.org/2021/12/08/dobbs-case-exposes-moral-bankruptcy-of-abortion-advocates/.

[111] Jamie Ehrlich, July 9th, 2020: https://edition.cnn.com/2020/06/29/politics/clarence-thomas-abortion-dissent/index.html.

[112] David Bjornstrom, Esq, "Life or Death" etc: https://personhood.org/2021/10/28/pro-life-and-pro-abortion-arguments-in-the-dobbs-case/.; cf. also Steven Ertelt, "Supreme Courts Lets" etc: https://www.lifenews.com/2021/12/10/

variability of who is elected when, whereas for those who are deliberately aborted, it is an irrevocable outcome for them.

But whatever comes of the current debate in America, this book has established the right to legal representation from conception on the basis that human rights are integral to human relationships; and, therefore, a human being is by definition a human being-in-relationship and, as such, entitled to the rights embodied in being human. Thus we need principles and practices that are not based on "Legal Positivism", the doctrine that a particular legal decision is *per se* right and irreformable just because it is embodied in a specific law. We need, rather, a principle of justice expressed in a law for the sake of all: non-discriminatory; irreformable; and adequate to all the needs of the unborn child.

Section ii: Viability is for Life. It is reported that 'viability' is roughly understood to be equivalent to 'when a

---

supreme-court-lets-texas-abortion-ban-keep-saving-babies-dismisses-joe-bidens-lawsuit-against-it/.

fetus can survive outside the womb'[113]. Thus there are a number of problems with the term 'viability'. Whether or not a child can survive being prematurely born or delivered is almost entirely dependent on the level of available technical equipment, expertise and the skill to manage the maturation of breathing and organ development. In other words, 'viability' is here understood to be dependant, or contingent upon, factors which have nothing to do with the inherent right of living in relationship to other human beings; indeed, just as a dependent has a right to be fed, clothed and educated, so the extreme dependency of a baby is even more a reason to respect the right to succour and support in his or her struggle to live.

'Viability', however, has a natural meaning of 'capable of living'[114]; indeed, as applied to the sperm and egg, viability means that 'pregnancy depends on the viability of

---

[113] Mark Sherman, "Supreme Court set to take up all-or-nothing abortion fight".

[114] Cf. "Viable": https://www.merriam-webster.com/dictionary/viable; and cf. "Fetal Viability": https://en.wikipedia.org/wiki/Fetal_viability.

the sperm and egg'[115]. In other words, although viability
tends to be used as a legal term, referring to the life of a
child being viable or capable of life outside of the
womb[116], being viable can clearly refer to the life of the
child from conception and his or her ongoing development in the womb. Being 'capable of life', then, really expresses the self-evident fact that being alive entails the capability of being alive: that being capable of life is an inherent and intrinsic property of the entity, whether it be
sperm, egg or fertilized human embryo. It is almost as if
the term, viability, could apply to the idea of *being alive in
such a way as to be able to continue to live.* Now, that being
the case, just as there are external factors that assist the
premature birth of a child to survive being premature, so
the natural environment of the human embryo is the
mother's womb: the place where, being alive in such a way
as to be able to continue to live, indicates an inherent

---

115 "Viability": https://www.google.com/search?q=viabil-
ity+definition&rlz=1C1SQJL_enGB840GB840&oq=Viabil-
ity+def&aqs=chrome.0.0i512j69i57j0i512l8.7340j0j7&sourceid
=chrome&ie=UTF-8.

116 "Fetal Viability": https://en.wikipedia.org/wiki/Fe-
tal_viability.

ability to live that is, at the same time, sustained by the natural or assisted environment of the child. In other words, 'viability' really refers to the existence of a child being alive of itself and, by definition, is alive both independently of the mother and of any assistance afforded the child once born. In other words, it is not that the child does not need the mother's womb to develop or that the child does not need specialist help if born prematurely; rather, viability refers to the fact of the life of the child is in itself independent of both mother and any subsequent support. Thus assisting what is already living is rooted, as it were, in the very fact of the human embryo being alive of itself. 'Viability', then, is evidenced in the very moment of conception, as we have seen, from which the embryological development proceeds of itself, *uninterruptedly,* from the first instant of fertilization until natural death.

Section iii: Choice, Burdens and their Alleviation. The problems of the burden of pregnancy and the burdens of parenting are not identical and are equally in need of addressing[117]; but, in either case, there is the question of

---

117 *Micaiah Bilger* | Dec 1 :

what choice has already been exercised and, in the event of pregnancy, the difference it makes of another life like that of ours.

With respect to the woman's choice

Justice Roberts says: "...if you think that the issue is one of choice, that women should have a choice to terminate their pregnancy, that supposes that there is a point at which they've had the fair choice, opportunity to [choose], and why would 15 weeks be an inappropriate line?' [118]

Clearly there is a variety of problems here: A woman may not have chosen pregnancy, either because of rape or a failed contraceptive or she has simply changed her mind, on becoming pregnant, and decided not to go through with being pregnant; but, even so, the question is no

---

2021: https://www.lifenews.com/2021/12/01/justice-amy-coney-barrett-destroys-pro-abortion-argument-that-abortion-bans-force-women-to-be-pregnant/.

[118] Jonah McKeown, December 2, 2021, « 'Dobbs v. Jackson'" etc: https://www.ncregister.com/cna/dobbs-v-jackson-what-did-roberts-kavanaugh-and-barrett-say.

longer that of a personal choice, as to whether to take the risk of contraceptives, or to become pregnant, or to end the life of a child that has come to exist, precisely because there is another human life that is implicated in the decision of whether or not to have a deliberate abortion. Who, as I say, represents the right of a child at risk if not an impartial authority whose task it is to safeguard the rights of all human beings, namely a court? In any other conflict of interest the child would be represented in court – Why not now, when, there is a risk to the child's very life? As regards the concept of 'fair choice' – how does that apply to the child? In the end, neither 'fair choice' or 'viability' are relevant if the court's role is to protect all innocent human life[119].

With respect to the claim that there is a burden of pregnancy, it has to be admitted that there are risks to the mother which, while they can be alleviated, cannot be totally excluded. On the one hand, there has been an

---

[119] Lauretta Brown, « After oral arguments" etc: https://www.ncregister.com/cna/after-oral-arguments-in-landmark-dobbs-v-jackson-abortion-case-experts-say-roe-s-days-are-numbered.

appreciation of mothers from ancient times; and, indeed, Tobias says to his son Tobit:

'Honor her all the days of your life; do what is pleasing to her, and do not grieve her. Remember, my son, that she faced many dangers for you while you were yet un-born' (Tobit, 4: 3-4). Let us not forget, either, that the Commandment of the Lord carries a promise: 'Honor your father and your mother, that your days may be long in the land which the Lord God gives you' (Ex 20: 12).

On the other hand, it is possible that gratitude for childbearing has diminished in a culture that often blames a mother for introducing another carbon footprint into the world.

However, the question is whether the risk to the mother is "equivalent" to a right to abortion". The risk to the mother of continuing the pregnancy is not equivalent to the risk to the child of the mother discontinuing the pregnancy; for, in reality, the child is at greater risk of dy-ing, especially if deliberately aborted, both because of be-ing taken out of his or her natural environment and be-cause of the means by which the child is removed. It is also true that there are many effects on women of what they do, ranging from the psycho-physical to the spiritual; even the so-called home abortion pill has a rising record of ill effects:

'As you may have seen in the media, a study [in the UK] has revealed that over 10,000 women have needed hospital treatment following the use of medical abortion pills since March 2020'[120].

In other words, there is a concrete burden on a woman of abortion itself. Similarly, any use of contraceptive hormones, never mind any other unnatural devices used to prevent or end a pregnancy, has an impact on the woman. Indeed, one of the most tragic effects of hormonal contraception, beyond the possible death of a child, is that it

---

[120] Cf. Email: Right To Life UK info@righttolife.org.uk, 06/12/2021: https://percuity.files.wordpress.com/2021/10/foi-ma-treatment-failure-211027.pdf; and cf. David Maddox, "Abortion Pill Horror": https://www.express.co.uk/life-style/health/1527888/Abortion-pill-diy-nhs-warning.

suppresses ovulation which is, apparently, a reliable indication of a woman's health[121].

As regards the burden of parenting, again it is realistic
to acknowledge that there are all kinds of difficulties for
parents, single or married, to address; however, we do so
in the context of many formative influences, whether of
our own parents, friends or the contribution of Church
groups. Having ten children in ten years, two of whom
were early miscarriages, it is clear that a variety of help is
needed. In the end, however, there are wider considerations, namely the infertility of others, the decline in the
birth-rate and, simply, the life of the child. Thus one of the
main provisions in American law, and there are probably

---

[121] Cf. Marguerite R. Duane and Erin Adams, "The State
of Fertility Awareness Based Method Education for Medical
Professionals," in *Humanae Vitae, 50 Years Later: Embracing
God's Vision for Marriage, Love, and Life*, ed. Theresa Notare
(Washington, DC: The Catholic University of America Press,
2019), p. 206: ovulation is monitored "since the presence of ovulation is a sign of good health" (and see footnote 40: Pilar Vigil
et al, *The Linacre Quarterly*, no. 4 2012); and Cf. Francis Etheredge, "A Touch of Experience: Where Are You?":
    https://www.hprweb.com/author/francis-etheredge/.

equivalents in other countries and cultures: are the 'safe haven laws, which allow mothers to relinquish their newborns to authorities without fear of repercussions, as an alternative to abortion and the burdens of motherhood'[122].

One final point, as regards the burdens of motherhood and parenting, is the implication of the following claim: 'Abortion advocates argue that legal abortion is necessary to give women equal rights because men do not get pregnant'[123]. Women's equal rights do not, however, depend on any denial of differences between men and women; rather, women's rights are precisely that: the rights of women as women. Rather, there is a woman's right to be helped with the burdens of motherhood and parenting, precisely because these are of their nature activities that entail a husband or a man's involvement. In other words,

---

[122] *Micaiah Bilger* | Dec 1 : "Justice Amy Coney Barrett" etc.

2021: https://www.lifenews.com/2021/12/01/justice-amy-coney-barrett etc.

[123] David Bjornstrom, Esq, "Life or Death" etc: https://personhood.org/2021/10/28/pro-life-and-pro-abortion-arguments-in-the-dobbs-case/.

even in the case of artificial insemination or other practices, the involvement of others implies the woman's right to help with the outcome of a woman becoming pregnant. Furthermore, if there is an act of injustice against the woman, how much more does she have a right to help? But, in the end, there is the injustice to the child, once conceived, if an abortion is proposed; and, therefore, the right of the child to assistance is an inalienable right: a right which cannot be disregarded without perpetrating a radical and irreversible injustice to the child.

Section iv: Bodily Integrity, Liberty, Equality and the Constitution. Although these points pertain to a specific case of American law, the features which are claimed to be relevant to the woman, all pertain to the child too: 'bodily integrity, liberty and equality'[124]. Once the sperm has entered the open orifice of the ovum, there is a new entity signified by a unilateral closing of all the external openings

---

[124] Steven Ertelt, 1ˢᵗ December, 2021: "Justice Clarence Thomas" etc.

https://www.lifenews.com/2021/12/01/justice-clarence-thomas-makes-it-clear-theres-no-right-to-abortion-in-the-constitution/.

in what is now the human embryo. In other words, the first instant of fertilization establishes the bodily integrity of the human embryo; and, by integrity, is meant the foundational wholeness of the new being, albeit it will differentiate into placenta and child in due course. Nevertheless, as the placenta is a necessary, albeit temporary organ regulating the nutrition and waste product requirements of the baby, it falls within the terms of bodily integrity. 'Liberty' pertains to the freedom from the life-threatening actions of others as indeed it pertains to the full development of the individual and his or her talents; for, without bodily existence, there is no possibility of taking advantage of life's opportunities as and when the child matures. 'Equality', too, is an expression of the human identity and dignity which is equal to that of the mother; and, dependent as the child is, this dependence does not detract from the radical equality of both mother and child being equally in receipt of the gift of life.

Further, there is the question, in the case of the American Constitution, of whether or not 'the Constitution is neutral on the question of abortion'[125]. But if, as was argued earlier, the Constitution was written in a time that presupposed the right to life and the presupposition that citizenship was established at birth, which again presupposes the right to life of the child-to-become-citizen, then this legal frame of reference implies both a right to life and the practical necessity of determining citizenship at the time of birth. At the same time, if historically it is clear that the Constitution is not neutral on abortion then it is also clear that either the American people have a referendum or make the Constitution, from this point on, neutral on abortion. However, neither a referendum nor making the Constitution neutral on abortion addresses the underlying question of the inherent right to life of an innocent human being, otherwise unrepresented legally and whose life is irreversibly endangered by an arbitrary decision of viability outside of the womb – when viability includes the very place where the child is naturally viable, namely the womb of the woman.

---

[125]   Justice   Kavanagh :   « Dobbs   v   Jackson"   etc.: https://www.ncregister.com/cna/dobbs-v-jackson-what-did-roberts-kavanaugh-and-barrett-say.

Section v: Brain Death and Abortion. Is it true that 'unborn babies are like brain dead people who lack the consciousness to feel pain'[126]? On the one hand, it is recognized that a baby is seen 'recoiling when being poked or touched'[127]; and, indeed, the skin of a human being is a natural expression of his or her psycho-physical integrity. But it is also true to say that the life of a baby expresses the progressive nature of human development and, if the child lost a limb, it would not cease to be a child and would continue to develop. If pain censors are characteristic of our skin, then their presence may well be both evident and active as a part of the normal reflex cycle of the nervous system; and, as such, pain will be present at the start of that reflex process. Thus a child losing a limb in the womb would undoubtedly be painful whether or not consciousness is explicitly "experienced". On the other hand, brain

---

126 Micaiah Bilger | Dec 2, 2021: « Doctor Slams Sotomayer" etc.

https://www.lifenews.com/2021/12/02/doctor-slams-sotomayor-to-compare-an-unborn-child-to-a-brain-dead-person-is-wholly-ignorant/.

127 Micaiah Bilger | Dec 2, 2021: « Doctor Slams Sotomayer" etc.

death is a controversial condition because it is often if not wholly linked to obtaining organs for transplants and the persons response to stimuli may simply be a sign that he or she is alive – albeit heavily sedated. Furthermore, those engaged in promoting death by assistance freely acknowledge the centrality of stopping the heart as a foundational criterion for death [128]. Similarly, those seeking to ensure the death of the child in the womb use the drug used to execute prisoners by injecting it directly into the child's heart [129]. In other words, while a dog is for life and would not be treated like this at all without social outrage,

---

[128] Cf. the forthcoming from *The Catholic Medical Quarterly*, UK, a three part article, "Loneliness, Euthanasia and the Wholeness of Human Personhood" which, among it many sources, quotes: "Doctors seek life-ending drugs that smooth the way for the terminally ill" by Lisa M. Krieger" etc: https://medicalxpress.com/news/2020-09-doctors-life-ending-drugs-smooth-terminally.html; and cf. also "Is an assisted death 'quick and painless'?" by Michael Cook:
https://alexschadenberg.blogspot.com/2021/11/is-assisted-death-quick-and-painless.html.

[129] Cf. Beattie, "Catholicism, Choice and Consciousness: A Feminist Theological Perspective on Abortion", p. 64.

a child is for eternal life and suffers immeasurably the rejection and destruction of an often inhospitable un-welcome.

Irrespective, however, of whether or not and how much an unborn child experiences pain; his or her death is an irreversible event; and, therefore, an act of discrimination[130] against the ongoing life and development of a child.

Section vi: On Abortion and the Advancement of Women. The argument that a child is an obstacle to the advancement of women: Do 'women need to abort their unborn babies to succeed'[131]? Justice Amy Barrett, a mother of seven children, and indeed many other working mothers, are clearly evidence that motherhood and work

---

[130] See the tendency in the UK to recognize the justice of providing for children with specials needs, in this case owing to Down's Syndrome: https://gript.ie/a-huge-achievement-groundbreaking-uk-down-syndrome-bill-passes-second-stage-in-parlia-ment/?mc_cid=e0a9ef2ad0&mc_eid=47b31ab45d.

[131] Cf/ *Micaiah Bilger* | Dec 1 : "Justice Amy Coney Barrett" etc.

and, more widely, fulfilment, are not incompatible. According to St. Edith Stein, the very opposite is the case in that the natural tendency of the woman to be sensitive to the whole of human personhood is precisely the gift to be preserved and contributed, whether domestically or internationally. St. Edith wrote:

'Indeed, no woman is only "woman"; every one has her individual gifts just as well as a man, and so is capable of professional work of one sort or another, whether it be artistic, scholarly, technical or any other. Theoretically this individual talent may extend to any sphere, even to those somewhat outside women's scope' .... 'When working out laws and decrees a man might perhaps aim at the most perfect legal form, with little regard to concrete situations; whereas a woman who remains faithful to her nature even in parliament or in the administrative services, will keep the concrete end in view and adapt the means accordingly'[132].

---

[132] From Edith Stein's "Essays on Woman" compiled from lectures given before she entered Carmel. This excerpt is here with permission and is translated from the German by Freda Mary Oben, Ph.D. Copyright© 1987, 1996 Washington

Section vii: True Justice is Irreversible. There is the argument that it is better not to reverse the law because it 'introduces uncertainty into the law' or, when people are regularly relying on it, to change the law becomes 'disruptive'[133] and therefore, in both cases, the law ceases to be a stable contributor to the necessary order that enables a society to function coherently. Should, then, the abolition of slavery never have taken place? Should allowing the "common man and woman" to vote not have been allowed? Should there be no correction of an unjust law and, therefore, no progress over time of the law expressing, more fully, the full recognition of justice before the law for all?

Province of Discalced Carmelites ICS Publications 2131 Lincoln Road, N.E. Washington, DC 20002-1199 U.S.A., quoted by Dr. Ronda Chervin, on pp. 65 and 75 of an unfinished manuscript supplied by the author, R. Chervin, of *The Battle for the 20ᵗʰ Century Mind* (Forthcoming from St. Luis, MO: En Route Books and Media, 2022).

[133] John McGuirk, 3/12/2021, "Is the Supreme Court" etc: https://gript.ie/is-the-us-supreme-court-really-about-to-ditch-roe-v-wade/?mc_cid=e0a9ef2ad0&mc_eid=47b31ab45d

However, by contrast, if laws are not corrected when recognized to be erroneous then injustice is perpetuated and the rule of law, by implication, called into contempt. Furthermore, justice being the natural objective of the law, any progress in the enactment of just laws or the correction of unjust ones, establishes hope for both these objectives in the future and, in the present, objectively improved conditions for those directly benefitting from the improvement in the law. Finally, the law is a teacher and, as such, has an obligation to teach what can be learnt concerning the right regulation of society for the benefit of all its members.

In the end, however, is it the place of the court to rule on the nature of human life or, rather, to recognize it as a foundational presupposition of justice for all?[134] Clearly, the whole reason that there is a persistent objection to the legalization of abortion is that it enacts an action against the life of human beings: the giving of which is non-discriminatory and is given by God unconditionally. Indeed, it could be argued, a just law is that which is non-discriminatory, unconditional and irrevocable in its recognition

---

[134] John McGuirk, 3/12/2021, "Is the Supreme Court" etc.

of the right to life of the innocent. In other words, human life needs the help of the protection of international law.

FURTHER READING
A VARIETY OF PRIOR WORK ON CONCEPTION

The argument advanced in this work is also explored in numerous articles and books, all of which entail their respective sources.

The two articles below were published by the Catholic Medical Quarterly, UK, and were also included in the book, ***Scripture: A Unique Word***: https://www.cambridgescholars.com/product/978-1-4438-6044-4:

Part I of II: "A Person from the first instant of Fertilization", *Catholic Medical Quarterly* **(August 2010, Vol. 60, No. 3, pp. 12-26)**

Part II of II: "A Person from the first instant of Fertilization", *Catholic Medical Quarterly,* **(November 2010, Vol. 60, No. 4, pp. 20-26)**

Chapter 7 of ***The Human Person: A Bioethical Word***: https://enroutebooksandmedia.com/bioethicalword/.

Chapter 5, Part II, of ***Conception: An Icon of the Beginning***, was written by two specialists: Professor Justo Aznar (a former member of the Pontifical Academy for Life) and Julio Tudela; this Chapter gives excellent evidence of the earliest interactions between the sperm and the ovum. In other words, no sooner do sperm and egg come into contact with one another than they start to interact and from then on everything is relevant to the formation of the human embryo: https://enroutebooksandmedia.com/conception/

As regards the theological argument, go to Chapter 5 of ***Mary and Bioethics: An Exploration***, this gives the reasoning that led to the view that Mary cannot be wholly holy unless there is an instantaneous union of the body and soul from the first instant that each exists and exists as one: https://enroutebooksandmedia.com/maryandbioethics/.

Articles published by the *National Catholic Bioethical Quarterly*:

"The Mysterious Instant of Conception", *National Catholic Bioethical Quarterly* of America, Vol. 12, Autumn 2012, No. 3, pp. 421-430.

"Frozen and Untouchable: A Double Injustice to the Embryo", *National Catholic Bioethics Quarterly* 16.1 (Spring 2016).

"The First Instant of Mary's Ensoulment", *National Catholic Bioethics Quarterly*, Vol. 19, Autumn 2019, pp. 359-367.

In general, I would like to thank all those who have collaborated over the years and, in many cases, continue to do so; for, in the end, while this work is my own, the extent to which it draws on the work of others is almost, indirectly, incalculable.

Printed in Great Britain
by Amazon

77537619R00122